# Lecture Notes in Biomathematics

Managing Editor: S. Levin

W0106868

**61**

# Resource Management

Proceedings of the Second Ralf Yorque Workshop
held in Ashland, Oregon, July 23–25, 1984

Edited by M. Mangel

Springer-Verlag
Berlin Heidelberg New York Tokyo

**Editor**

Marc Mangel
Department of Mathematics, University of California
Davis, CA 95616 USA

ISBN-13: 978-3-540-15982-7      e-ISBN-13: 978-3-642-46562-8
DOI: 10.1007/978-3-642-46562-8

2146/3140-543210

## Preface

These are the proceedings of the Second R. Yorque Workshop on Resource
Management which took place in Ashland, Oregon on July 23-25, 1984. The
purpose of the workshop is to provide an informal atmosphere for the
discussion of resource assessment and management problems. Each participant
presented a one hour morning talk; afternoons were reserved for informal
chatting. The workshop was successful in stimulating ideas and interaction.

The papers by R. Deriso, R. Hilborn and C. Walters all address the same
basic issue, so they are lumped together. Other than that, the order to the
papers in this volume was determined in the same fashion as the order of
speakers during the workshop -- by random draw.

Marc Mangel
Department of Mathematics
University of California
Davis, California
June 1985

# TABLE OF CONTENTS

# A GENERAL THEORY FOR FISHERY MODELING

Jon Schnute
Pacific Biological Station
Nanaimo, B.C. V9R 5K6
Canada

## INTRODUCTION

Is fisheries management a science? Here is one answer, given by Larkin (1972): "In brief, our fisheries literature is largely unscientific in the strict sense of the word, and our fisheries management is unscientific in almost every sense of the word." In this paper I describe a mathematical theory which, I believe, significantly improves the possibilities for scientific management. To explain my basis for this claim, I must begin with a discussion of science itself, and I have chosen a historical example from a completely different field.

Over a decade ago I had occasion to ponder the nature of science when, as a summer project, I produced a manuscript on the history of man's investigations into the vibrating string (Schnute 1973). Some of the earliest work dates back to Pythagoras (c. 580-500 B.C), who discovered that the vibratory pitch is inversely proportional to the string length. This idea achieved quasi-mathematical proof in the work of the Dutch scholar Isaac Beeckman (1570-1637), who used an argument based on similar triangles to show that particles on a string would need to move twice as far as comparable particles on another string of half the length. Later, the French Jesuit Marin Mersenne (1588-1648) made the first recorded auditory observations of the many harmonics within a single vibrating string. Galileo (1564-1642) introduced empirical measurement into these philosophical speculations by using iron chisel marks on a brass plate as a device for counting the actual frequencies of vibration. Christiaan Huygens (1629-1695) gave highly mathematical descriptions of the pendulum and the vibration of a single weight on a string. Isaac Newton (1643-1727) put a major component of theory in place by formulating his famous general law that forces and accelerations are proportional, with mass as the constant of proportionality. Later writers, such as Brook Taylor (1685-1731), John Bernoulli (1667-1748), Daniel Bernoulli (1700-1782), and Jean D'Alembert (1717-1783) applied the physics of the time to various configurations of weighted strings, either swinging like a pendulum or vibrating with both ends fixed. Finally, Leornard Euler (1708-1783) consolidated much of this work by formalizing Newton's ideas into a general "principle of mechanics", which involved writing down second order differential equations based on the forces, masses, and accelerations appropriate to the particular situation. His explicit solution for the vibrating string appeared in 1749. Ten years later, in 1759, Joseph-Louis Lagrange (1736-1813) at the age of 23 published a spectacular paper showing explicitly that a string loaded with N weights has N harmonics, and that as N becomes large the solutions approach the infinite set of harmonics in a continuous string. Even after this definitive stroke of genius, it took at least until the late 1800s for all the mathematcal arguments to be polished into a state of unequivocal rigor.

So what is scientific progress? The example above illustrates that it is an evolutionary pursuit of the truth as an increasingly comprehensive theory is developed to include previous knowledge and speculation. This evolutionary process never really ends; at any point in time, a new perspective can lead to a new synthesis of past understandings. When such a perspective emerges, it makes the field more scientific in the sense that the relationship of parts to the whole is better understood.

Although the principles of scientific development just described are generally well known among the scientific community (indeed they seem almost like motherhood statements), I claim that fisheries science is now ripe to apply them in a fashion new to this discipline. By way of analogy with the historical example above, I would place us in the period between Newton and Euler, when various specialized techniques were applied to specific cases of vibration. For example, consider the theories for analysis of catch and effort data. We have, at an early stage, the two methods of Schaefer (1954, 1957) and their refinement by Pella and Tomlinson (1967). Later Schnute (1977) put these ideas on a more solid analytic and statistical footing, but conceptually the basic ideas remained much the same. Deriso (1980) accomplished a more dramatic refinement by showing that a suitable model could be obtained from biological first principles, provided that the resulting equations are summed over the age groups within the population. On the other hand, Roff (1983) seemed to get better practical results with a simple "black box" model, independent of biological theory.

If these diverse ideas are to coalesce into a coherent pattern, then clearly a general theory is required, just as it was for the various special solutions to the problem of vibration. Unfortunately for the theorist, fisheries are complex phenomena, not readily amenable to a universal description. Box and Jenkins (1970), in their definitive text on time series analysis, suggested a general method for modeling complex systems. Their strategy (1970, pp. 18-19) involves four steps, paraphrased below:

Step 1. From the interaction of theory and practice, postulate a useful class of models appropriate to the situation. A typical model in the class may potentially involve a large number of parameters.

Step 2. Use data and a knowledge of the system to identify a particular tentative model from the class. This involves restricting model parameters in a meaningful way.

Step 3. Estimate the parameters in the tentative model of step 2.

Step 4. Perform diagnostic checking. Is the tentative model adequate? If not, return to step 2 and revise accordingly. If so, use the model for forecasting and control.

A key concept in this strategy is the class of models mentioned in step 1. If we are to choose among a group of possible models, then we must be able to compare the candidates; consequently, the candidates themselves must all be special cases of some general class. For example, Box and Jenkins proposed the class of ARMA (autoregressive moving average) models for time series. Their step 2 involves deciding how many lagged terms should be included in the autoregression and moving average. Revision at step 4 involves either increasing or reducing the number of terms in the model. At the end of the process, all models entertained are comparable because they fit into the unified class postulated in step 1.

The strategy of Box and Jenkins, then, is a technique for developing and applying general theory. Their strategy, however, is not to be confused with the particular class of ARMA models that they advocate. In the present context, ARMA models are inadequate for at

least two reasons. First, they are not based on biological theory and
cannot, therefore, be used in conjunction with biological information
beyond the data set to be investigated. Second, they are linear, while
biological processes, such as recruitment, sometimes are not.

In this paper, I derive a class of models for analysing catch and
effort data, based on assumptions of the Deriso type. (See Schnute,
1985, for further details.) The class has some features in common with
the AR models of Box and Jenkins; however, it is based on traditional
biological principles of growth, mortality, and recruitment.
Furthermore, it turns out to include the models of Schaefer (1957),
Pella and Tomlinson (1969), Walter (1973), Schnute (1977), Deriso
(1980), and Roff (1983) as special cases. Thus the class accomplishes
my stated scientific purpose of consolidating much earlier work.

The ideas given here have practical implications for the technique
of fishery modeling. The modeler does not simply apply a universal
model for all catch and effort data. Instead, he or she (step 1) uses a
general framework for constructing such models, with enough
flexibility to accomodate a wide range of biological assumptions. The
modeler's task is then (step 2) to identify those particular
assumptions that apply to the case at hand, (step 3) to fit these into
the general framework, and thus to build an appropriate model for the
stock in question. Since biological assumptions are often tenuous, it
may be reasonable (step 4) to build several related models and to
compare the results. The advantage of a universal framework is that the
various results are then comparable, since all are special cases of a
larger whole.


## 1. THE MODEL

### 1A. Notation and assumptions

A consistent notation is important for stating the model and its
consequences properly, particularly since the model involves numerous
mathematical objects. There are two basic measures of time (in years):
age a and year t. It is assumed that there is a fixed age k at which
fish are recruited to the fishery. Deriso (1980) calls this
"knife-edged" recruitment. In order not to detract from a few central
features of the model, I have devoted this paper entirely to the
knife-edged case, although a more general "incomplete recruitment"
model similar to that discussed by Deriso could also be developed.

Thus, in this paper a fish of age a in year t was born in year t-a and recruited in year t-a+k.

The model involves three main functions of a and t: the total number of available fish (the population) $N_{at}$ aged a at the start of year t, the number $C_{at}$ caught at age a during year t, and the weight $w_{at}$ of a single fish aged a in year t. Corresponding to the numbers $N_{at}$ and $C_{at}$ are the biomasses

$$(1.1) \qquad N_{at}^* = w_{at} N_{at} \, , \qquad C_{at}^* = w_{at} C_{at} \, .$$

Throughout this paper, a superscript asterisk distinguishes biomasses of fish from numbers of fish. The model also involves two functions of t alone: the recruitment $R_t$ of fish of age k at the start of year t, and the fishing effort $E_t$ during year t. Again, there is a related biomass

$$(1.2) \qquad R_t^* = w_{kt} R_t \, .$$

Furthermore, corresponding to age structured populations, are the totals

$$(1.3) \qquad N_t = \sum_{a=k}^{\infty} N_{at} \, , \qquad C_t = \sum_{a=k}^{\infty} C_{at} \, ,$$

and related biomass totals

$$(1.4) \qquad N_t^* = \sum_{a=k}^{\infty} N_{at}^* \, , \qquad C_t^* = \sum_{a=k}^{\infty} C_{at}^* \, .$$

Definitions (1.3)-(1.4) reflect a convention used throughout this paper: when the first subscript a is dropped from a doubly subscripted quantity like $N_{at}$, the resulting function of t refers to the sum over recruited ages.

In addition to the above functions of a and t, there are numerous parameters which drive the model. One of these is the recruitment (age k) weight $V_t$ associated with the cohort born in year t and recruited in year t+k. This same cohort is assumed to have an extrapolated pre-recruitment (age k-1) weight $v_t$, and a Ford growth coefficient $\rho$. By definition of $v_t$ and $V_t$

$$(1.5) \qquad v_t = w_{k-1,t+k-1} \, ; \qquad V_t = w_{k,t+k} \, .$$

In (1.5) the pre-recruitment weight $v_t$ should be interpreted as a parameter, rather than the acutal weight at age k-1; for this reason it is called an "extrapolated" weight. The model simply uses $v_t$, $V_t$ and ρ as three parameters to define the growth curve for ages a≥k. Typically, fish younger than age k have weights that do not conform to the mature growth curve. The model also involves a mortality parameter $M_t$ and a catchability parameter $q_t$ associated with natural mortality and fishing, respectively. These allow the definition of three fractions: natural survival $\sigma_t$, fishing survival $\phi_t$, and total survival $\tau_t$. Finally, there are three parameters $\alpha_t$, β, and γ related to recruitment. All this notation is summarized in Tables 1 and 2. Various parameters, such as $M_t$, are presumed to depend on t, rather than being constants, to allow stochastic behavior to be introduced systematically in the model. For example, variation in natural mortality can be accounted for by assuming that $M_t$ is a random variable with mean M. Similarly, $q_t$, $V_t$, and $\alpha_t$ can be used to introduce random fluctuations in catchability, growth, and recruitment, respectively.

Table 1. Notation for the primary quantities which define the model (1.6)-(1.13). A dimensionless quantity is indicated by a "-" in the units column.

| Notation | Meaning | Typical units |
|---|---|---|
| | Time | |
| a | fish age | years |
| t | year of the fishery | years |
| | Functions | |
| $N_{at}$ | population of age a fish at the start of year t | fish |
| $C_{at}$ | catch of age a fish during year t | fish |
| $w_{at}$ | weight of a fish at age a in year t | kg |
| $R_t$ | recruitment at the start of year t | fish |
| $E_t$ | fishing effort during year t | boat-days |
| | Parameters | |
| k | age of recruitment | years |
| $M_t$ | instantaneous natural mortality in year t | - |
| $q_t$ | catchability in year t | 1/(boat-days) |
| $v_t$ | pre-recruitment weight of a fish born in year t | kg |
| $V_t$ | recruitment weight of a fish born in year t | kg |
| ρ | Ford's growth coefficient | - |
| $\alpha_t$ | recruitment productivity parameter | fish/kg |
| β | recruitment optimality parameter | 1/kg |
| γ | recruitment limitation parameter | - |

TABLE 2. Notation defined in terms of the model's primary quantities listed in Table 1. The number of the defining equation from the text is shown, along with typical units of measurement.

| Not'n | Eq'n | Meaning | Units |
|---|---|---|---|
| | | Functions | |
| $N^*_{at}$ | (1.1) | population biomass at age a, start of year t | kg |
| $C^*_{at}$ | (1.1) | biomass at age a, caught in year t | kg |
| $N_t$ | (1.3) | total available population, start of year t | fish |
| $N^*_t$ | (1.4) | total available biomass, start of year t | kg |
| $C_t$ | (1.3) | number of fish caught during year t | fish |
| $C^*_t$ | (1.4) | biomass of fish caught during year t | kg |
| $R^*_t$ | (1.2) | biomass of fish recruited at the start of year t | kg |
| | | Parameters | |
| $\phi_t$ | (1.12) | survival from fishing mortality in year t | - |
| $\sigma_t$ | (1.11) | survival from natural mortality during year t | - |
| $\tau_t$ | (1.10) | survival from all mortality during year t | - |
| $W_t$ | (1.17) | asymptotic weight of a fish born in year t | kg |
| $K$ | (1.18) | Brody's growth coefficient | 1/yr |
| $a_{o,t}$ | (1.19) | extrapolated age of 0 weight for birth year t | yr |
| $R_m$ | (1.29) | maximum possible recruitment | fish |
| $S_m$ | (1.29) | stock for maximum recruitment | fish |
| $S_o$ | (1.30) | carrying capacity | fish |

Using the above notation, the model consists of eight assumptions, listed below:

(1.6) $\quad N_{a+1,t+1} = \tau_t N_{at}$ ,

(1.7) $\quad C_{at} = \lceil 1-\phi_t \rceil N_{at}$ ,

(1.8) $\quad N_{kt} = R_t \{ N^*_{t-k} - C^*_{t-k} \}$ ,

(1.9) $\quad w_{a+1,t+1} - w_{at} = \rho(w_{at} - w_{a-1,t-1})$ ,

(1.10) $\quad \tau_t = \sigma_t \phi_t$ ,

(1.11a) $\quad \sigma_t = \exp(-M_t)$ ,

(1.12a)   $\phi_t = \exp(-q_t E_t)$   ,

(1.13a)   $R_t\{S\} = \alpha_t S(1-\beta\gamma S)^{1/\gamma}$   .

The first three equations describe fish population dynamics, and the fourth describes growth dynamics. The next three give the particulars of survival from total, natural, and fishing mortalities, respectively. Finally, the last equation details the recruitment process. Equations (1.6), (1.7), and (1.9) all apply only to recruited ages $a \geq k$, with the understanding in (1.9) that, when $a=k$, the pre-recruitment weight $w_{k-1,t-1}$ is extrapolated. In (1.8) and (1.13a), braces designate functional dependence; thus "$R_t\{S\}$" means that $R_t$ is a function of S.

The model is constructed so that the last three equations can readily be modified without altering some of the main conclusions. For example, three alternatives to the pair (1.11a)-(1.12a) are:

(1.11b)   $\sigma_t = \dfrac{M_t + q_t E_t}{M_t \exp(M_t + q_t E_t) + q_t E_t}$   ,

(1.12b)   $\phi_t = \dfrac{M_t + q_t E_t \exp(-M_t - q_t E_t)}{M_t + q_t E_t}$   ;

or

(1.11c)   $\sigma_t = \{1 + \exp(q_t E_t)[\exp(M_t)-1]\}^{-1}$   ,

(1.12c)   $\phi_t = 1 - \exp(-M_t)[1 - \exp(-q_t E_t)]$   ;

or

(1.11d)   $\sigma_t = \exp(-M_t)$   ,

(1.12d)   $\phi_t = 1 - q_t E_t$   .

An alternative to (2.13a) is:

(1.13b)   $R_t = R$   ,

where R is a constant. The significance of each model assumption, including the various alternatives, is described in the paragraphs following.


1B. Population dynamics


The first assumption (1.6) summarizes the process of aging one year. For age $a \geq k$, the cohort population $N_{at}$ at the start of year t is simply multiplied by the total survival fraction $\tau_t$ to obtain number of fish remaining in this cohort at the start of the next year. Similarly, if $\phi_t$ represents the fraction of the population which survives fishing, then (1.7) expresses catch $C_{at}$ as the complementary fraction $1-\phi_t$ of the cohort population. The most important aspect of both these assumptions is that survival does not depend on age after the recruitment age k. This will be true if both natural and fishing mortality are age-independent. Although constant natural mortality is a rather common assumption, the requirement that catchability should not vary with age is more severe. As mentioned earlier, the model assumes that fishing gear has knife-edged selectivity for recruited vs. non-recruited fish. Thus, the model does not include a more general selection ogive (Beverton and Holt 1957, p. 75-82; Fournier and Archibald 1982, p. 1198, equations 2.1 and 2.2), a restriction which is perhaps reasonable for the model's intended use on catch data without age information.

The third equation (1.8) describes recruitment of age k fish in year t as a function of the post-fishing biomass in year t-k. The particular form of the function $R_t\{S\}$, where S is the stock biomass, is given in (1.13), one of the flexible assumptions identified earlier. The most important aspects of (1.8) are, first, that recruitment occurs at age k and, second, that recruitment depends on available biomass. If, for example, recruitment were to depend instead on available population (remove the two asterisks from (1.8)), then some of the model's main conclusions would change, as explained later.


1C. Growth details


The fourth model assumption (1.9) is essentially Ford's (1933) growth model (Ricker 1975, p. 222, equation 9.17). Ricker refers to $\rho$ as "Ford's growth coefficient". Typically $\rho$ is a fraction, $0 < \rho < 1$; thus (1.9) states that the annual weight increment decreases by the factor $\rho$ each year. Values of $\rho \geq 1$ are also possible, and the growth increment is constant when $\rho=1$.  When $\rho \neq 1$, it follows that

$$(1.14) \qquad w_{at} = v_{t-a} + (V_{t-a}-v_{t-a}) \; \frac{1-\rho^{1+a-k}}{1-\rho} \qquad .$$

This is obtained by solving the difference equation (1.9), using the initial conditions (1.5) rewritten with different indices as:

$$(1.15) \qquad w_{k-1,t-1} = v_{t-k} \; ; \qquad w_{kt} = V_{t-k} \qquad .$$

The reader can verify that (1.14) satisfies both conditions (1.15), and a proof follows by mathematical induction. Consistent with earlier discussion, (1.14) applies only to recruited ages $a \geq k$; the pre-recruitment weight $v_{t-a}$ is merely a convenient parameter.

The growth equation (1.14) can be written in the alternative form:

$$(1.16) \qquad w_{at} = W_{t-a}[1 - e^{-K(a-a_{o,t-a})}] \qquad ,$$

where

$$(1.17) \qquad W_{t-a} = (V_{t-a}-\rho v_{t-a})/(1-\rho) \qquad ,$$

$$(1.18) \qquad K = -\log \rho \qquad ,$$

and

$$(1.19) \qquad a_{o,t-a} = k-1 - \frac{\log[(V_{t-a}-v_{t-a})/(V_{t-a}-\rho v_{t-a})]}{\log \rho} \qquad .$$

Here (1.16) is the familiar "von Bertalanffy growth curve" with three parameters (1.17)-(1.19) interpretted, respectively, as the extrapolated asymptotic weight W of a fish born in year t-a, Brody's growth coefficient K (Ricker 1975, p. 221), and the extrapolated age $a_o$ when a fish born in year t-a has weight 0.

The two growth curves (1.14) and (1.16) are equivalent with different parameters as follows:

| Equation | Parameters |
|----------|------------|
| (1.14)   | v, V, $\rho$ |
| (1.16)   | W, K, $a_0$ |

Although (1.16) is the most common form of the curve, (1.14) is preferable here for two reasons. First, equations of the general theory below take the simplest form when (1.14) is used. Second, the growth model adopted by Deriso (1980, equation 1), is (1.14) with the restriction

$$(1.20) \qquad v = 0 \qquad .$$

From (1.19), (1.20) is equivalent to the requirement

$$(1.21) \qquad a_0 = k-1$$

in the von Bertalanffy equation (1.16). Thus Deriso's growth model extrapolates back to weight 0 at the pre-recruit age k-1.


1D. Mortality details

The fifth model assumption (1.10) states that the total survival fraction is the product of survivals from fishing and natural mortality. Given the earlier equations involving $\tau_t$ and $\phi_t$, this apparently reasonable assumption really amounts to a definition of $\sigma_t$. A short algebraic calculation based on (1.6), (1.7), and (1.10) shows that

$$(1.22) \qquad N_{a+1,t+1} = \sigma_t(N_{at} - C_{at}) \qquad ;$$

that is, $\sigma_t$ represents the fraction of fish not caught in the fishery that survives to the next year. Fishing and natural mortality may occur concurrently; however, from (1.7) $\phi_t$ is the fraction of the original population that survives the fishery, and from (1.24) $\sigma_t$ is the fraction of the remaining population that survives natural mortality.

The model is flexible as to the precise details whereby fishing and natural mortality occur. If fishing occurs first, in a pulse immediately following recruitment, then (1.11a)-(1.12a) describe the relevant survivals based on mortality $M_t$, catchability $q_t$, and

effort $E_t$. If fishing and natural mortality taken place
simultaneously throughout the year (as assummed, for example, by
Beverton and Holt 1957; Ricker 1975; Fournier and Archibald 1982;
Kimura and Tagart 1982), then (1.11b)-(1.12b) apply. If pulse fishing
occurs at the end of the year, just prior to recruitment, then
(1.11c)-(1.12c) are correct. All three of these formula pairs have the
common feature that the product $\sigma_t \phi_t$, the total mortality, in
each case is

$$(1.23) \qquad \tau_t = e^{-(M_t + q_t E_t)} \quad .$$

## 1E. Recruitment details

The final model assumption (1.13a) is essentially the recruitment
function proposed by Deriso (1980); however, it is written here with
somewhat different parameters to allow greater flexibility. Specific
values of the parameter $\gamma$ determine four cases of special interest,
namely

$$(1.24) \qquad \gamma = -\infty: \quad R_t\{S\} = \alpha_t S \quad ;$$

$$(1.25) \qquad \gamma = -1: \quad R_t\{S\} = \alpha_t S/(1+\beta S) \quad ;$$

$$(1.26) \qquad \gamma = 0: \quad R_t\{S\} = \alpha_t S \, e^{-\beta S} \quad ;$$

$$(1.27) \qquad \gamma = 1: \quad R_t\{S\} = \alpha_t S(1-\beta S) \quad .$$

Each case above is either a valid mathematical limit for (1.13a) or
simply the result of substituting the given value of $\gamma$. The first case
(1.24) is a constant productivity model; $\alpha_t$ recruits are produced
for each unit of stock biomass, regardless of stock size. The next
three cases (1.25)-(1.27) correspond to classical recruitment models,
according to the following table:

| Equation | Reference |
|----------|-----------|
| (1.25) | Beverton and Holt (1957, p. 49, equation 6.10) |
| (1.26) | Ricker (1954; 1958, p. 238, equation 11.7) |
| (1.27) | Schaefer (1954, bottom of p. 34) |

To complete the dicussion of (1.13a), it is convenient to drop the subscripts:

(1.28)  $R = \alpha S(1-\beta\gamma S)^{1/\gamma}$  .

The parameters $\alpha$ and $\beta$ should always be positive, although $\gamma$ can have any sign. The curve (1.28) in the SR-plane always passes through the origin (0,0). When $\gamma>-1$, it has a maximum point $(S_m, R_m)$ with coordinates

(1.29)  $S_m = [\beta(1+\gamma)]^{-1}$; $R_m = \alpha\beta^{-1}(1+\gamma)^{-(1+\gamma)/\gamma}$  .

Furthermore, when $\gamma>0$, the curve also has a second point $S_0$ (in addition to $S=0$) where $R=0$, namely

(1.30)  $S_0 = (\beta\gamma)^{-1}$  .

From (1.29)-(1.30), it follows that

(1.31)  $S_m = [\gamma/(1+\gamma)] S_0$  .

When $\gamma=1$, the factor $\gamma/(1+\gamma)$ is 1/2, and the high point of the curve is mid-way between the intecepts at 0 and $S_0$. This fact corresponds to the symmetry of the Schaefer curve. In that context the quantity $S_0$ is called the "carrying capacity" (Schnute 1977), and the stock size $S_m$ for optimum recruitment is well-known to be half the carrying capacity. To avoid this possibly unrealistic symmetry, Pella and Tomlinson (1969) added a parameter to the Schaefer model. Here the factor $\gamma/(1+\gamma)$ serves essentially the same purpose, as shown in (1.31).


## 2. DIFFERENCE EQUATIONS

The previous section completely describes an age structured model for a fish stock. It consists of the eight assumptions (1.6)-(1.13). These are quite general, although their main limitations are, first, that fishing and natural mortalities are presumed age-independent and, second, that recruitment to the fishery occurs at a distinct age k. To apply this model to total annual catch and effort data, it is necessary to derive equations that involve these data only. Such equations are called difference equations, because they typically involve changes, or differences, in functions of time t. In particular, this section proves mathematically and explains biologically the following eight consequences of the first five assumptions (1.6)-(1.10):

$$(2.1) \qquad N_{a+1,t+1} - \tau_t N_{at} = 0 \quad,$$

$$(2.2) \qquad N^*_{a+1,t+1} - \tau_t N^*_{at} = \rho \tau_t (N^*_{at} - \tau_{t-1} N^*_{a-1,t-1}) \quad,$$

$$(2.3) \qquad N_{t+1} - \tau_t N_t - R_{t+1} = 0 \quad,$$

$$(2.4) \qquad N^*_{t+1} - \tau_t N^*_t - R^*_{t+1} = \rho \tau_t (N^*_t - \tau_{t-1} N^*_{t-1} - v_{t-k} R^*_t / V_{t-k}) \quad,$$

$$(2.5) \qquad N_t = C_t / (1-\phi_t) \quad,$$

$$(2.6) \qquad N^*_t = C^*_t / (1-\phi_t) \quad,$$

$$(2.7) \qquad N^*_{t+1} = (1+\rho) \tau_t N^*_t - \rho \tau_t \tau_{t-1} N^*_{t-1}$$

$$+ V_{t+1-k} R_{t+1} \{\phi_{t+1-k} N^*_{t+1-k}\} - \rho \tau_t v_{t-k} R_t \{\phi_{t-k} N^*_{t-k}\} \quad,$$

$$(2.8) \qquad C^*_{t+1}/(1-\phi_{t+1}) = (1+\rho) \tau_t C^*_t/(1-\phi_t) - \rho \tau_t \tau_{t-1} C^*_{t-1}/(1-\phi_{t-1})$$

$$+ V_{t+1-k} R_{t+1} \{\phi_{t+1-k} C^*_{t+1-k}/(1-\phi_{t+1-k})\}$$

$$- \rho \tau_t v_{t-k} R_t \{\phi_{t-k} C^*_{t-k}/(1-\phi_{t-k})\} \quad.$$

Equation (2.1) is a minor variation of assumption (1.6). It can be regarded as a conservation statement; those fish at age a that survive year t are precisely the fish at age a+1 that begin the next year t+1. Similarly, (2.2) is a related conservation principle for biomass; however, the principle is not quite so simple because, due to growth, the weights $w_{at}$ and $w_{a+1,t+1}$ are not the same. To prove (2.2), notice first from (1.6) that for $a \geq k+1$

$$N_{a+1,t+1} = \tau_t N_{at} = \tau_t \tau_{t-1} N_{a-1,t-1} \quad.$$

Subscripts of N in each of these three equal terms match the various subscripts of w in (1.9). Multiplying the above terms by their counterparts in (1.9) gives

$$N^*_{a+1,t+1} - \tau_t N^*_{at} = \rho(\tau_t N^*_{at} - \tau_t \tau_{t-1} N^*_{a-1,t-1}) \quad ,$$

that is, (2.2). Notice that (2.2) and (1.9) are very similar mathematically, particularly when $\tau_t$ and $\tau_{t-1}$ are both 1. Biologically, (2.2) expresses conservation, not of biomass itself, but rather of biomass increments.

Equation (2.3) is similar to (2.1), except that (2.3) describes the entire population, rather than just one age class. Again, (2.3) is a conservation priciple; the population in year t+1 consists of survivors and recruits. To prove (2.3), sum (2.1) over ages a from k to ∞, and notice that

$$\sum_{a=k}^{\infty} N_{a+1,t+1} = \sum_{a=k+1}^{\infty} N_{a,t+1} = N_{t+1} - R_{t+1} \quad .$$

This completes the proof of (2.3).

Just as (2.3) is the counterpart of (2.1) for the total population, (2.4) is the counterpart of (2.2) for the total biomass. Furthermore, the proof is similar. Since (2.2) is valid for $a \geq k+1$, the proof begins by summing (2.2) over ages from k+1 to ∞. Notice that

$$\sum_{a=k+1}^{\infty} N^*_{a+1,t+1} = N^*_{t+1} - N^*_{k+1,t+1} - N^*_{k,t+1}$$

$$= N^*_{t+1} - w_{k+1,t+1} \tau_t R_t - w_{k,t+1} R_{t+1} \quad ,$$

$$\sum_{a=k+1}^{\infty} N^*_{at} = N^*_t - w_{kt} R_t \quad ,$$

and

$$\sum_{a=k+1}^{\infty} N^*_{a-1,t-1} = N^*_{t-1} \quad .$$

Applying these three results to the sum of (2.2) and collecting terms gives

(2.9)     $N^*_{t+1} - \tau_t N^*_t - w_{k,t+1} R_{t+1}$

$$= \rho \tau_t (N^*_t - \tau_{t-1} N^*_{t-1} - [w_{kt} + \rho^{-1}(w_{kt} - w_{k+1,t+1})] R_t) \quad .$$

Here the expression in square brackets reduces simply to $w_{k-1,t-1}$ as a result of the growth equation (1.9) when a=k. Consequently, (2.9) becomes (2.4), given the definitions (1.2) and (1.5) for $R^*$, v, and V. This completes the proof of (2.4).

Like (2.2), (2.4) expresses conservation of biomass increments, in this case for the total population. Equation (2.4) is the counterpart of a similar equation in Deriso (1980a, p. 269, equation 4), except that the earlier equation excludes the final term involving $v_{t-k}$ because Deriso's growth model has the parameter $v_t = 0$ for every t. The new term gives (2.4) an interesting symmetry in the role of $R^*$ on each side of the equation.

Equation (2.5) is a simple consequence of (1.7); it follows as usual on summing over ages a from k to ∞. Since $\phi_t$ represents survival, $1-\phi_t$ refers to mortality. Thus $C_t/(1-\phi_t)$ might be called the "catch per unit mortality", CPUM, in comparison with the more conventional catch per unit effort, CPUE. Even though CPUE may not index population in this model, (2.5) shows that the CPUM does. Similarly, (2.6) shows that the same concept applies to biomass. The straightforward proof is based on multiplying (1.7) by $w_{at}$ and summing over ages a.

It follows from (2.6) that

$$N^*_t - C^*_t = \phi_t N^*_t \quad ;$$

consequently, from (1.2), (1.15), and (1.8),

(2.10)     $R^*_t = V_{t-k} R_t \{\phi_{t-k} N^*_{t-k}\} \quad ,$

where it is understood that the function $R_t\{.\}$ is specified by an assumption like (1.13a) or (1.13b). By combining (2.4) and (2.10), one obtains (2.7), which is a recursion formula for calculating the biomass $N^*_{t+1}$ at time t+1 based on the biomass at the previous times t, t-1, t+1-k, and t-k. Furthermore, (2.6) allows one to convert (2.7) into (2.8), a similar recursion formula for the catch $C_{t+1}^*$. Even though (2.7) and (2.8) look rather complex, their biological meaning is identical to (2.4), that is, a conservation equation for biomass increments.

Equation (2.8) satisfies the criterion mentioned at the start of this section: it involves only the catch and effort data, $C_t^*$ and $E_t$, for various years t. The dependence on $C_t^*$ is shown explicitly, and $E_t$ enters the equation through $\tau_t$ and $\phi_t$ via (1.10) and one of the formula pairs (1.11)-(1.12). Practical application of (2.8) involves making specific assumptions about the various parameters in the equation.  For example, suppose that all parameters are independent of time and that survival and recruitment are given by (1.11a), (1.12a), and (1.13a). Then (2.8) becomes

$$(2.11) \qquad (1-e^{-qE_{t+1}})^{-1} C_{t+1}^* =$$

$$(1+\rho)e^{-M}(e^{qE_t}-1)^{-1} C_t^* - \rho e^{-2M-qE_t}(e^{qE_{t-1}}-1)^{-1} C_{t-1}^*$$

$$+ \alpha V(e^{qE_{t+1-k}}-1)^{-1} C_{t+1-k}^* [1-\beta\gamma(e^{qE_{t+1-k}}-1)^{-1} C_{t+1-k}^*]^{1/\gamma}$$

$$- \rho\alpha v e^{-M-qE_t}(e^{qE_{t-k}}-1)^{-1} C_{t-k}^* [1-\beta\gamma(e^{qE_{t-k}}-1)^{-1} C_{t-k}^*]^{1/\gamma} \quad .$$

Although (2.11) is rather complex, it can be regarded as a predictive equation for $C_{t+1}^*$ based on the data:

$$(2.12) \qquad C_{t-k}^*, \ C_{t+1-k}^*, \ C_{t-1}^*, \ C_t^*, \ E_{t-k}, \ E_{t+1-k}, \ E_{t-1}, \ E_t, \ E_{t+1},$$

and the seven parameters:

$$(2.13) \qquad M, \ q, \ \rho, \ \alpha v, \ \alpha V, \ \beta, \ \gamma,$$

where the eighth parameter k is presumed known. (Here $\alpha$, v, and V are confounded because these parameters occur only in the products $\alpha v$ and $\alpha V$.) In principle, then, (2.11) could be used to estimate the parameters (2.13) by, say, nonlinear regression.

For many data sets, the estimation problem suggested by
(2.11)-(2.13) may not be completely solvable; also, there may be
problems related to the type of statistical error in (2.11). However,
this example illustrates a general method to construct a model for
catch and effort data. The key equation is (2.8), called the <u>catch
equation</u> in the rest of this paper. This follows from the first five
model assumptions (1.6)-(1.10). The modeler's initial task is to
incorporate into the catch equation a set of additional assumptions
appropriate to the fishery in question. These always concern the
parameters which drive the model. For example, (2.11) is based on
time-independent parameters, as well as the last three model
assumptions (1.11a), (1.12a), and(1.13a). This is only of a very large
number of possibilities. Other models might be based on
(1.11c)-(1.12c). Furthermore, the parameters might be allowed to
depend on time; for example, survivals might be related to
environmental observations. Some parameters, such as those associated
with growth and mortality, might be known from survey data. Fishing
mortality might be independent of effort, or, as in (1.13b),
recruitment might be independent of stock biomass. The modeler must
apply specific knowledge of the fishery to tailor the model to the
situation. The advantage of the method here is that it provides a
single framework in which numerous models can be tried and compared
easily with one another. Thus the catch equation serves as a starting
point for the iterative model-building technique of Box and Jenkins
(1970); it specifies a useful <u>class</u> of fisheries models. The next
section compiles some of the possibilities.


## 3. SPECIAL CASES

The model (1.6)-(1.13) involves three flexible biological
components: mortality, growth, and recruitment. To obtain an orderly
list of variations in the catch equation, it is necessary first to
define the alternatives available for each of these three components.
In this section, consider the case of constant parameters discussed
above in connection with (2.11). Table 3 lists some of the main ways in
which the model can be specialized. There are two options for
mortality, three for growth, and four for recuitment, with an
additional set of four possible recruitment submodels. All these cases
have been singled out because they fundamentally affect the number of
time lags and parameters in the catch equation, as explained later. In
the theory here, these assumptions serve to guide the biologist in

identifying a tentative model, along the lines of "step 2" in Box and
Jenkins' (1970) model building strategy. The next few paragraphs
describe in detail the options listed in Table 3.

TABLE 3. Special cases of the general model, corresponding to
particular assumptions on mortality, growth, and recruitment. Each case
is assigned a label and a brief description. The model parameters for
each case are also listed.

| Label | Parameters | Defining assumptions | Description |
|-------|-----------|---------------------|-------------|
| | | Mortality | |
| M1 | M, q | (1.11a) and (1.12a) | General mortality model |
| M2 | $\tau$, $\phi$ | (1.11a) and $\phi$=constant | Constant fishing mortality |
| | | Growth | |
| G1 | $\rho$, V, v | (1.14) with v≠V | General growth model |
| G2 | $\rho$, V | (1.14) with v=0 | Deriso's growth model |
| G3 | V | (1.14) with v=V | Constant weight |
| | | Recruitment | |
| R1 | $\alpha$, $\beta$, $\gamma$ | (1.13a) with k>2 | General recruitment model |
| R2 | $\alpha$, $\beta$, $\gamma$ | (1.13a) with k=2 | Two-year recruitment |
| R3 | $\alpha$, $\beta$, $\gamma$ | (1.13a) with k=1 | One-year recruitment |
| R4 | R | (1.13b) | Constant recruitment |
| | | Recruitment submodels for i=1, 2, or 3 | |
| Ria | $\alpha$, $\beta$ | Model Ri with $\gamma$=1 | Schaefer recruitment |
| Rib | $\alpha$, $\beta$ | Model Ri with $\gamma$=0 | Ricker recruitment |
| Ric | $\alpha$, $\beta$ | Model Ri with $\gamma$=-1 | Beverton-Holt recruitment |
| Rid | $\alpha$ | Model Ri with $\beta$=0 | Constant productivity |

3A. Mortality components

Model M1 is simply the general case (1.11a)-(1.12a) in which
fishing occurs at the start of the year; the associated parameters are
M and q. The alternatives (1.11b)-(1.12b) and (1.11c)-(1.12c) are not
particularly important here, as they do not fundamentally affect the
structure of the catch equation. A more significant alternative is M2,
the assumption that fishing mortality is constant. This model would be
valid in fisheries where effort is adjusted by fishermen or by
regulation to take a fixed annual fraction of available stock. For

example, effort might be held at a fixed level from year to year,
regardless of stock size; or effort might increase in years of low
stock, but with a corresponding loss of efficiency. The resulting
version of (2.8),

$$(3.1) \qquad C^*_{t+1} = (1+\rho)\tau C^*_{t+1} - \rho\tau^2 C^*_{t-1}$$

$$+ V(1-\phi)R_{t+1}\{\phi C^*_{t+1-k}/(1-\phi)\}$$

$$- \rho\tau v(1-\phi)R_t\{\phi C^*_{t-k}/(1-\phi)\} \quad ,$$

involves only the catch C*, not the effort $E_t$. Here $\phi$ and $\tau$ are
used as the associated mortality parameters.


3B.  Growth components

Model G1 is the general case (1.14), with parameters $\rho$, V, and v.
G2 is Deriso's (1980) restriction (1.20) that the extrapolated
pre-recruitment weight v is 0; this leaves only the two parameters $\rho$
and V. G3 is the case of constant weight

$$(3.2) \qquad v = V \quad ;$$

by (1.14), (3.2) implies that $w_{at}$=V for all a and t. In this case the
value of $\rho$ is irrelevant in (1.14); furthermore, if

$$(3.3) \qquad \rho = 0 \quad ,$$

then the difference equation (1.9) also implies constant weight.  Thus
(3.2) and (3.3) are companion conditions; either implies constant
weight for ages a$\geq$k. When (3.3) is true, the catch equation (2.8)
assumes the much simpler form

$$(3.4) \qquad C^*_{t+1}/(1-\phi_{t+1}) = \tau_t C^*_t/(1-\phi_t)$$

$$+ VR_{t+1}\{\phi_{t+1-k}C^*_{t+1-k}/(1-\phi_{t+1-k})\} \quad .$$

This equation can also be obtained another way. When the weight $w_{at}$ is constant, there is no essential mathematical distinction between $N_t^*$ and $N_t$; these quantities differ only by the fixed factor $V$. Similar remarks apply to $R_t^*$ and $R_t$. Consequently, the result (2.3) is valid with superscript asterisks on $N$ and $R$; call this revised version (2.3*). Then (3.4) follows from (2.3*) exactly as the earlier equation (2.8) follows from (2.4). Notice that (2.3*) is much simpler than (2.4); (2.3*) is a first order difference equation, while (2.4) is second order. Biologically, the assumption of constant weight allows conservation of biomass increment to be replaced by conservation of biomass itself. Indeed, (3.4) is just a biomass conservation equation.

The constant weight equation (3.4) has one other potential application. Suppose that recruitment depends on population numbers rather than biomass and that the catch data are in numbers rather than weight. Mathematically, one can represent this situation simply by letting $V=1$ and dropping the superscript asterisks. Then (3.4) becomes

$$(3.5) \qquad C_{t+1}/(1-\phi_{t+1}) = \tau_t C_t/(1-\phi_t) + R_{t+1} \qquad ,$$

which is the catch equation for $C_t$ (as opposed to $C^*$) when recruitment depends on population numbers or, as in (1.13b), is independent of population variables.

## 3C. Recruitment components

The model $R_1$ is simply the general recruitment model (1.13a) with parameters $\alpha$, $\beta$, and $\gamma$. The case $k>2$ ($R_1$) is distinguished from $k=2$ and $k=1$ (R2 and R3, respectively) because the recruitment age $k$ influences the lags appearing in the catch equation. Model R4 is the constant recruitment case (1.13b). In this case the catch equation (2.8) reduces to

$$(3.6) \qquad C_{t+1}^*/(1-\phi_{t+1}) = (1+\rho)\tau_t C_t^*/(1-\phi_t)$$

$$- \rho\tau_t\tau_{t-1}C_{t-1}^*/(1-\phi_{t-1}) + (V-\rho v\tau_t)R \qquad .$$

Finally, models R1, R2, and R3 all have submodels corresponding to special values of $\gamma$, as shown in Table 3. For example, R1a is Schaefer recruitment with k>2, and R3b is Ricker recruitment with k=1.


## 3D. Combined components


The two mortality models, three growth models, and four recruitment models give a total of 24 (2 x 3 x 4) possible variations to the catch equation. These are listed in Table 4, along with the corresponding model parameters and lags in catch and effort used to predict $C_{t+1}^{*}$. (For example, $C_t^{*}$ has lag 1.) The full set of 24 possible equations need not all be listed here, but Table 5 shows all those obtained by combinations of (1) M1 or M2, (2) G1 or G3, and (3) R1 or R4. In general, the catch equation for a model follows from (2.8) or its counterpart (2.11) with constant parameters. The cases M2, G3, and R4 can be obtained from the specialized versions of the catch equation (3.1), (3.4), and (3.6), respectively.

TABLE 4. Models obtained by combining assumptions on mortality, growth, and recruitment. Labels are consistent with Table 3. Each model is listed with its associated catch lags, effort lags (if any), parameters, and parameter count.

| MODEL | | | LAGS | | PARAMETERS | |
|---|---|---|---|---|---|---|
| M | G | R | Catch lags | Effort lags | Parameter list | Count |
| 1 | 1 | 1 | 1,2,k,k+1 | 0,1,2,k,k+1 | $M, q, \rho, \alpha V, \alpha v, \beta, \gamma$ | 7 |
| 1 | 1 | 2 | 1,2,3 | 0,1,2,3 | $M, q, \rho, \alpha V, \alpha v, \beta, \gamma$ | 7 |
| 1 | 1 | 3 | 1,2 | 0,1,2 | $M, q, \rho, \alpha V, \alpha v, \beta, \gamma$ | 7 |
| 1 | 1 | 4 | 1,2 | 0,1,2 | $M, q, \rho, RV, Rv$ | 5 |
| 1 | 2 | 1 | 1,2,k | 0,1,2,k | $M, q, \rho, \alpha V, \beta, \gamma$ | 6 |
| 1 | 2 | 2 | 1,2 | 0,1,2 | $M, q, \rho, \alpha V, \beta, \gamma$ | 6 |
| 1 | 2 | 3 | 1,2 | 0,1,2 | $M, q, \rho, \alpha V, \beta, \gamma$ | 6 |
| 1 | 2 | 4 | 1,2 | 0,1,2 | $M, q, \rho, RV$ | 4 |
| 1 | 3 | 1 | 1,k | 0,1,k | $M, q, \alpha V, \beta, \gamma$ | 5 |
| 1 | 3 | 2 | 1,2 | 0,1,2 | $M, q, \alpha V, \beta, \gamma$ | 5 |
| 1 | 3 | 3 | 1 | 0,1 | $M, q, \alpha V, \beta, \gamma$ | 5 |
| 1 | 3 | 4 | 1 | 0,1 | $M, q, RV$ | 3 |
| 2 | 1 | 1 | 1,2,k,k+1 | – | $\tau, \rho, \phi \alpha V, \phi \alpha v, \phi \beta /(1-\phi), \gamma$ | 6 |
| 2 | 1 | 2 | 1,2,3 | – | $\tau, \rho, \phi \alpha V, \phi \alpha v, \phi \beta /(1-\phi), \gamma$ | 6 |
| 2 | 1 | 3 | 1,2 | – | $\tau, \rho, \phi \alpha V, \phi \alpha v, \phi \beta /(1-\phi), \gamma$ | 6 |
| 2 | 1 | 4 | 1,2 | – | $\tau, \rho, (1-\phi)R(V-\rho \tau v)$ | 3 |
| 2 | 2 | 1 | 1,2,k | – | $\tau, \rho, \phi \alpha V, \phi \beta /(1-\phi), \gamma$ | 5 |
| 2 | 2 | 2 | 1,2 | – | $\tau, \rho, \phi \alpha V, \phi \beta /(1-\phi), \gamma$ | 5 |
| 2 | 2 | 3 | 1,2 | – | $\tau, \rho, \phi \alpha V, \phi \beta /(1-\phi), \gamma$ | 5 |
| 2 | 2 | 4 | 1,2 | – | $\tau, \rho, (1-\phi)RV$ | 3 |
| 2 | 3 | 1 | 1,k | – | $\tau, \phi \alpha V, \phi \beta /(1-\phi), \gamma$ | 4 |
| 2 | 3 | 2 | 1,2 | – | $\tau, \phi \alpha V, \phi \beta /(1-\phi), \gamma$ | 4 |
| 2 | 3 | 3 | 1 | – | $\tau, \phi \alpha V, \phi \beta /(1-\phi), \gamma$ | 4 |
| 2 | 3 | 4 | 1 | – | $\tau, (1-\phi)RV$ | 2 |

TABLE 5. All possible catch equations for the combined mortality models M1 or M2, growth models G1 or G3, and recruitment models R1 or R4. Model numbers are labelled as the triple <MGR>.

<111>
$$(1-e^{-qE_{t+1}})^{-1}C^*_{t+1} =$$

$$(1+\rho)e^{-M}(e^{qE_t}-1)^{-1}C^*_t - \rho e^{-2M-qE_t}(e^{qE_{t-1}}-1)^{-1}C^*_{t-1}$$

$$+ \alpha V(e^{qE_{t+1-k}}-1)^{-1}C^*_{t+1-k}[1-\beta\gamma(e^{qE_{t+1-k}}-1)^{-1}C^*_{t+1-k}]^{1/\gamma}$$

$$- \rho\alpha v e^{-M-qE_t}(e^{qE_{t-k}}-1)^{-1}C^*_{t-k}[1-\beta\gamma(e^{qE_{t-k}}-1)^{-1}C^*_{t-k}]^{1/\gamma} .$$

<114>
$$\frac{C^*_{t+1}}{1-e^{-qE_{t+1}}} = \frac{(1+\rho)e^{-M}C^*_t}{e^{qE_t}-1} - \frac{\rho e^{-2M-qE_t}C^*_{t-1}}{e^{qE_{t-1}}-1} + RV - Rv\rho e^{-M-qE_t}$$

<131>
$$(1-e^{-qE_{t+1}})^{-1}C^*_{t+1} = e^{-M}(e^{qE_t}-1)^{-1}C^*_t$$

$$+\alpha V(e^{qE_{t+1-k}}-1)^{-1}C^*_{t+1-k}[1-\beta\gamma(e^{qE_{t+1-k}}-1)^{-1}C^*_{t+1-k}]^{1/\gamma}$$

<134>
$$\frac{C^*_{t+1}}{1-e^{-qE_{t+1}}} = \frac{e^{-M}C^*_t}{e^{qE_t}-1} + RV$$

<211>
$$C^*_{t+1} = (1+\rho)\tau C^*_t - \rho\tau^2 C^*_{t-1}$$

$$+ \alpha V\phi C^*_{t+1-k}[1 - \beta\gamma\phi(1-\phi)^{-1}C^*_{t+1-k}]^{1/\gamma}$$

$$- \rho\alpha v\tau\phi C^*_{t-k}[1 - \beta\gamma\phi(1-\phi)^{-1}C^*_{t-k}]^{1/\gamma}$$

<214> $\qquad$ $C^*_{t+1} = (1+\rho)\tau C^*_t - \rho\tau^2 C^*_{t-1} + (1-\phi)RV - (1-\phi)\rho\tau Rv$

<231> $\qquad$ $C^*_{t+1} = \tau C^*_t + \alpha V\phi C^*_{t+1-k}[1 - \beta\gamma\phi(1-\phi)^{-1}C^*_{t+1-k}]^{1/\gamma}$

<234> $\qquad$ $C^*_{t+1} = \tau C^*_t + (1-\phi)RV$

It is convenient to devise a compact notation to refer to a model. Let "<mgr>" refer to the combination of components Mm, Gg, and Rr. For example, <111> is the full model (2.11) with seven parameters (2.13). Similarly, <214> is the constant effort and recruitment model obtained by setting $R_t$ constant in (3.1), or, equivalently, setting $\tau_t$ and $\phi_t$ constant in (3.6). Various historical models can, at least apporoximately, be represented in this notation as follows:

| Model | Historical reference |
|-------|---------------------|
| <133a> | Schaefer (1957), Schnute (1977) |
| <133> | Pella and Tomlinson (1969) |
| <121> | Deriso (1980) |
| <134> | Roff (1983) |

In any model <mgr>, it is important to take note of the number of estimable parameters. As Table 4 indicates, this is not always the complete list of parameters for the separate model components in Table 3. For example, the full model <111> theoretically involves eight parameters in Table 3: M, q, $\rho$, V, v, $\alpha$, $\beta$, and $\gamma$. However, Table 4 (and the earlier discussion of (2.11)-(2.13)) shows that the three parameters $\alpha$, V, and v are confounded, so that <111> involves only seven parameters. This is not surprising biologically. If only the catch biomass $C_t$ is available, the model does not know if there are a large number of small fish or a smaller number of large fish. Consequently, the number recruited (proportional to $\alpha$) is confounded with fish weight (proportional to V and v). As the model becomes simpler, the list of estimable parameters becomes smaller. For example, <234> involves only two parameters: $\tau$ and $(1-\phi)RV$. Again this is resonable biologically. Since the model assumes constant fishing mortality, weight, and recruitment, it cannot distinguish among fishing survival $1-\phi$, weight V, and recruitment numbers R; raising the value of one of these parameters can be compensated by lowering another.

The practitioner should anticipate problems with confounded parameters when attempting to find estimates from actual data. In real applications, some parameters are usually available independently and should not be estimated from catch and effort data. This is particularly true for the growth parameters $\rho$, $V$, and $v$. A compelling advantage to the approach here is that it allows the modeler to build on prior biological knowledge.

## 4. DISCUSSION

The essence of the theory here is a concise set of biological assumptions from which a catch equation, involving only observed catches and efforts, can be derived. The assumptions encompass a spectrum of traditional hypotheses about growth, mortality, and recruitment; and the catch equation can be specialized in ways consistent with most historical models for catch and effort data. Nothing is said here about methods of introducing stochastic error in the model and estimating the model's parameters. Also no method is given for identifying the particular special case of the catch equation that is optimal for a given situation. These matters are discussed further, but still rather inadequately, by Schnute (1985). Even for Box and Jenkins (1970), model identification is a rather informal process, and speculation on methods of identifying ARMA models continues to this day.

When scientists hit upon a consolidating idea like the one presented here, it is perhaps common for them to get rather excited and speak of the importance of theory and fundamental methodology. Perhaps they may (as I have done) weave their own names into introductions that also mention real geniuses like Newton, Euler, Darwin, Einstein, or Fisher. I hope the reader will excuse me if I have been overly dramatic. It is, however, my honest opinion that fisheries science is somewhat adrift with a burgeoning literature on special techniques for handling this or that particular problem. The practitioner is continually faced with a vast menu of possible methodologies and a lurking sense that there might be still others more contemporary, but hidden in obscure journals or "grey" literature. As I see it, the only hope for the scientific community, faced with such a morass, is to begin moving in a direction with enough commonality among researchers that it becomes possible to compare their results easily and objectively. This would seem to require an agreed general theory. After all, it has happened in other fields; for example, researchers in electomagnetics at least have Maxwell's equations as one possible starting point. I realize that we can never have such tidy models as Maxwell's equations in fisheries, but this paper shows that there is at least some hope for a central theory. I don't know if this is the right one, but I do think that real progress is impossible without something like it.

## REFERENCES

Beverton, R. J. H., and S. J. Holt. 1957. On the dynamics of exploited fish populations. U. K. Min. Agric. Fish. Food, Fish. Invest. (Ser. 2) 19:   533p.

Box, G. E. P., and G. M. Jenkins. 1970. Time series analysis forecasting and control. Holden-Day. San Francisco, CA. xix + 553 p.

Deriso, R. B. 1980. Harvesting strategies and parameter estimation for an age-structured model. Can. J. Fish. Aquat. Sci. 37: 268-282.

Ford, E.  1933.  An account of the herring investigations conducted at Plymouth during the years from 1924-1933.  J. Mar. Biol. Assoc. UK. 19:   305-384.

Fournier, D. and C. P. Archibald. 1982. A general theory for analyzing catch at age data. Can. J. Fish. Aquat. Sci. 39: 1195-1207.

Kimura, D. K., and J. V. Tagart. 1982. Stock Reduction Analysis, another solution to the catch equations. Can. J. Fish. Aquat. Sci. 39: 1467-1472.

Larkin, P. A.  1972.  A confidential memorandum on fisheries science. Chapter 12 in "World Fisheries Policy:  A multidisciplinary View." B. J. Rothschild, ed. University of Washington Press, Seattle.

Pella, J. J., and P. K. Tomlinson.  1969.  A generalized stock production model.  Inter.-Am. Trop. Tuna Comm. Bull. 13:   421-496.

Roff, D. A. 1983. Analysis of catch/effort data: a comparison of three methods. Can. J. Fish. Aquat. Sci. 40: 1496-1506.

Ricker, W. E.  1954.  Stock and recruitment.  J. Fish. Res. Board Can. 11:   559-623.

      1958. Handbook of computations for biological statistics of fish populations. Bull. Fish. Res. Board Can. 119:   300 p.

      1975. Computation and interpretation of biological statistics of fish populations. Fish. Res. Board Can. Bull. 191:   xviii + 382 p.

Schaefer, M. B. 1954. Some aspects of the dynamics of populations important to the management of the commercial marine fisheries. Inter.-Am. Trop. Tuna Comm. Bull. 1: 25-56.

      1957. A study of the dynamics of the fishery for yellowfin
tuna in the eastern tropical Pacific Ocean. Inter.-Am. Trop. Tuna
Comm. Bull. 2: 245-285.

Schnute, J. 1973. Vibrating strings: the history of a metaphor.
    Dept. of Mathematics, University of B.C. 151 p. + 51 plates
    (Unpublished internal report).

      1977. Improved estimates from the Schaefer production model:
theoretical considerations. J. Fish. Res. Board Can. 34:  583-603.

      1985.  A general theory for analysis of catch and effort
data.  Can. J. Fish. Aquat. Sci. 42:  414-429.

Walter, G. G. 1973. Delay-differential equation models for fisheries.
    J. Fish. Res. Board Can. 30: 939-945.

DATA TRANSFORMATIONS IN REGRESSION ANALYSIS WITH APPLICATIONS TO STOCK - RECRUITMENT
RELATIONSHIPS

David Ruppert

and

Raymond J. Carroll

Department of Statistics
University of North Carolina
Chapel Hill, NC 27514/USA

## Abstract

We propose a methodology for fitting theoretical models to data. The dependent
variable (or response) and the model are transformed in the same way. Two types of
transformations, power transformation and weighting, are used together to remove
skewness and to induce constant variance. Our method is applied to the stock-
recruitment data of four fish stocks. Also discussed are estimates of the
conditional mean and the conditional quantiles of the original response.

## 1. Introduction

In regression analysis one seeks to establish a relationship between a response
y and independent variables $\underline{x} = (x_1,\ldots,x_k)'$. Often the physical or biological
system generating the data suggest that in the absense of random error $y = f(\underline{x},\underline{\theta})$
where f is a known function and $\underline{\theta}$ is an unknown parameter. In reality random
variability, modeling errors, and measurement errors will cause y to deviate from
$f(\underline{x},\underline{\theta})$.

Usually $\underline{\theta}$ is estimated by the (possibly nonlinear) least-squares estimator $\hat{\underline{\theta}}$
which minimizes

$$\sum_{t=1}^{N} (Y_t - f(\underline{x}_t,\hat{\underline{\theta}}))^2$$

where $y_t$ and $\underline{x}_t = (x_{1t},\ldots,x_{kt})'$ are the t-th observations of the response and the
independent variables, $t=1,\ldots,N$. The method of least squares is not uniquely
determined in the following sense. If h(y) is a monotonic function then $y = f(x,\theta)$
implies that $h(y) = h(f(\underline{x},\theta))$. In the absense of random error, there is an infinity
of possible models, one for each h. Taking h(y) to be the new dependent variable
and $h(f(\underline{x},\theta))$ to be the new regression model, the least-squares estimate $\hat{\theta}(h)$,
depending on h, minimizes

$$\sum_{t=1}^{N} \{h(y_t) - h(f(\underline{x}_t, \hat{\underline{\theta}}))\}^2 .$$

This paper addresses the question "how should h be chosen?" In the past h was often chosen so that the model was linear in the parameters, but with the wide availability of nonlinear-least squares software linearization is no longer necessary. Instead h should be selected so that $\hat{\theta}(h)$ has high statistical efficiency in some sense, say low asymptotic mean square error. For nonlinear models finite-sample mean square errors are seldom known and large-sample, or asymptotic, mean square errors must be used.

Least-squares estimation is asymptotically efficient when the errors are additive, normally distributed, and homoscedastic, that is, if

$$y_t = f(\underline{x}_t, \underline{\theta}) + e_t , \tag{1}$$

where $e_1, \ldots, e_N$ are independent $N(0, \sigma^2)$ random variables for some $\sigma^2 > 0$. If (1) fails to hold for the original response and model, it may hold after these have been transformed. For example, if the errors are multiplicative and lognormal, then

$$\log(y_i) = \log(f(\underline{x}_t, \underline{\theta})) + e_t$$

so that $h(y) = \log(y)$ is the appropriate transformation.

In general, it is impossible to know a priori how the random errors affect y. It is, however, often reasonable to assume as an approximation that for an unknown h

$$h(y_t) = h(f(\underline{x}_t, \theta)) + e_t \text{ and } e_t \sim N(0, \sigma^2),$$

and then to estimate h from the data.

Carroll and Ruppert (1984) introduced a methodology where h is assumed to belong to a parametric class such as the class of power functions. Specifically, $h(y) = h(y, \lambda)$ where $h(y, \lambda)$ is a known parametric family and $\lambda$ is an unknown parameter. Then $\underline{\theta}$, $\sigma$, and $\lambda$ are estimated simultaneously, for example by maximum likelihood.

Box and Cox (1964) used a modified power transformation family

$$y^{(\lambda)} = (y^\lambda - 1)/\lambda \qquad \lambda \neq 0$$
$$= \log(y) \qquad \lambda = 0,$$

which includes the log transformation in a natural, i.e., continuous, fashion. It should be mentioned that Box and Cox were concerned with a quite different transformation methodology. They transform only the response, not the model; see Carroll and Ruppert (1984).

Another transformation family which, as we will see, is useful in stock-recruitment analysis is division by $(u_t)^\alpha$ where $u_t$ is a known constant and $\alpha$ is an unknown parameter. Since $u_t$ is a constant which does not depend on $y_t$, a transformation of this kind has no effect on skewness but it induces homoscedasticity if $(u_t)^\alpha$ is the standard deviation of $y_t$. The constant $u_t$ can be one of the independent variables, a variable not depending on $\underline{x}_t$ or $y_t$, or a function of such variables. Division by $(u_t)^\alpha$ will be called a "power weighting transformation". Note that $u_t^\alpha$ denotes ordinary, not Box-Cox, power transformation of $u_t$ so that $u_t^\alpha = 1$ when $\alpha = 0$.

This paper is restricted to a model combining a Box-Cox power transformation with a power weighting transformation:

$$y_t^{(\lambda)} / u_t^\alpha = f^{(\lambda)}(\underline{x}_t, \underline{\theta})/u_t^\alpha + e_i. \tag{2}$$

However, the methodology that is developed here can be applied to other transformation families.

According by model (2), $y_t^{(\lambda)}$ is symmetrically distributed about $f^{(\lambda)}(\underline{x}_t, \underline{\theta})$. Therefore, $f^{(\lambda)}(\underline{x}_t, \underline{\theta})$ gives both the conditional (given $\underline{x}_i$) mean and median of $y_t^{(\lambda)}$. This implies further that the untransformed model $f(\underline{x}_t, \underline{\theta})$ gives the conditional median of the untransformed response $y_t$. However, the conditional mean of $y_t$ is not $f(\underline{x}_t, \underline{\theta})$ except if $\lambda = 1$. The conditional mean is discussed below. We see then that $f(\underline{x}_t, \underline{\theta})$ has two interpretations; it gives the value of $y_t$ if there is no error, and otherwise it gives the conditional median of $y_t$.

In our examples, $\underline{x}_t = x_{1t}$ is the size of a spawning stock $S_t$ and the response $y_t$ is the total return $R_t$. Commonly used model functions $f(\underline{x}, \underline{\theta})$ include the Ricker (1954) model

$$R_t = S_t \exp(\theta_1 + \theta_2 S_t) , \tag{3}$$

and the Beverton-Holt (1957) model

$$R_t = 1/(\theta_1 + \theta_2/S_t). \tag{4}$$

We will let $u_t = X_t$. Then (3) can be transformed to

$$R_t/S_t = \exp(\theta_1 + \theta_2 S_t) \tag{5}$$

by letting $\lambda=1$ and $\alpha=1$. By using $\lambda=0$ and $\alpha=0$ on (5), one obtains

$$\log(R_t/S_t) = \theta_1 + \theta_2 S_t. \tag{6}$$

Equation (6) is often favored since it is linear in the parameters.

Equation (4) can be linearized to

$$1/R_t = \theta_1 + \theta_2 \ (1/S_t) \tag{7}$$

or

$$S_t/R_t = \theta_2 + \theta_1 \ S_t \tag{8}$$

corresponding to $\lambda=-1$ and $\alpha=0$ and $\lambda-1$ and $\alpha=-1$, respectively.

By estimating $\lambda$ and $\alpha$ we can tell when these linearizing transformations are appropriate for a given set of data, and we can find more suitable transformations when the linearization is not appropriate.

Although we advocate transforming the response to achieve an error structure where least-squares is efficient, often one must estimate characteristics (such as the conditional mean) of the original response. In section 5 estimates of the conditional mean and the conditional quantiles of $y_t$ given $\underline{x}_t$ are presented.

## 2. Power transformations, skewness, and heteroscedasticity

In this section, we discuss how power transformations affect, and in particular remove, skewness and heteroscedasticity. Suppose that the random variable $y_t$ has mean $m_t$ and variance $\sigma_t^2 = g(m_t)$; the same function g applies for each t. If $y_t$ is transformed to $h(y_t)$, then by a Taylor approximation as in Bartlett (1947), the variance of $h(y_t)$ is

$$E\{h(y_t) - E\ h(y_t)\}^2 \doteq$$

$$E\{h(y_t) - h(m_t)\}^2$$

$$\doteq (\dot{h}(m_t))^2 \ E\{y_t - m_t\}^2$$

$$= (\dot{h}(m_t))^2 \ g(m_t)$$

where $\dot{h}(y) = d/dt\ h(y)$. The variance of $h(y_t)$ is approximately constant if $\dot{h}(y)$ is proportional to $g^{-\frac{1}{2}}(y)$.

In many cases g is a power function. For example, $g(m) = m$ for Poisson-distributed data and $g(m)$ is proportion to $m^2$ if the coefficient of variation (CV) is constant. Also, $g(m) = m^2$ for exponentially distributed data. When $g(m) \propto m^{2(1-\lambda)}$ then $\dot{h}(m) \propto g^{-\frac{1}{2}}(m)$ if $h(y) = y^{(\lambda)}$. In the case $\lambda = 0$ (constant CV), the log transformation stabilizes the variance, and for Poisson data, $\lambda = 1/2$, the square-root transformation is indicated.

The effect of h on skewness is also revealed by $\dot{h}$. Suppose that y is positively skewed. The extended right tail in the distribution of y is reduced through transformation to h(y) if large values of y are "compressed together" more than small values; more precisely, if for all $\triangle > 0$, $|h(y_1+\triangle) - h(y_1)| < |h(y_2+\triangle) - h(y_2)|$ whenever $y_1 > y_2$. Such compression occurs if $h(y)$ is a decreasing function, in which case h is called concave.

When $h(y)$ is a Box-Cox power transformation then $\dot{h}(y) = (d/dy) \ \lambda^{-1}(y^\lambda-1)$ when $\lambda\neq0$ or $\dot{h}(y) = (d/dy) \log y$ when $\lambda = 0$. In either case, $\dot{h}(y) = y^{\lambda-1}$, and consequently h is concave if $\lambda \leq 1$. Box-Cox transformations reduce right-skewness when $\lambda < 1$ and the amount of reduction increases as $\lambda$ decreases.

Left skewness if reduced by transformations that are convex, that is, which have an increasing derivative. If $\lambda \geq 1$, then $y^{(\lambda)}$ is convex.

## 3. Simultaneous power transformation and weighting by maximum likelihood.

Although power transformations can reduce both skewness and heteroscedasticity, the value of $\lambda$ inducing normality, or at least a reasonably symmetric distribution, need not be the same $\lambda$ which transforms to constant variance. A more flexible approach to modeling combines a Box-Cox power transformation to normality with a power weighting transformation to homoscedasticity as in equation (2). In this section we discuss the estimation of $\lambda$, $\alpha$, $\theta$, and $\sigma$ by maximum likelihood and the construction of a confidence region for $\lambda$ and $\alpha$ by likelihood-ratio testing.

To find the likelihood, we first note that the density of $e_t$ is

$$(2\pi \ \sigma^2)^{-\frac{1}{2}} \ \exp(-e^2/(2 \ \sigma^2))$$

and the Jacobian of $e_t \rightarrow y_t^{(\lambda)}$ is $y_t^{\lambda-1}/u_t^\alpha$. Therefore, the conditional density of $y_t$ given $x_t$ and $u_t$ is

$$f(y_t|x_t,u_t,\theta,\sigma,\alpha,\lambda) =$$

$$(2\pi \ \sigma^2)^{-\frac{1}{2}} \ (y_t^{\lambda-1}/u_t^\alpha) \ \exp(-[y_t^{(\lambda)} - f^{(\lambda)}(\underline{x}_t,\underline{\theta})]^2/(2 \ \sigma^2 \ u_t^{2\alpha})).$$

If $x_t$ and $u_t$ are fixed constants, not depending on $y_1,\ldots,y_{t-1}$, then the log-likelihood of $y_1,\ldots,y_N$ is

$$L(\lambda,\alpha,\theta,\sigma^2) = -N/2 \ \log(2\pi \ \sigma^2)$$

$$+ \ (\lambda-1) \sum_{t=1}^{N} \log(y_t) - \alpha \sum_{t=1}^{N} \log(u_t)$$

$$- \ (1/2 \ \sigma^2) \sum_{t=1}^{N} \{y_t^{(\lambda)} - f^{(\lambda)}(\underline{x}_t,\underline{\theta})\}^2/u_t^{2\alpha} .$$

If $x_t$ and $u_t$ depend on previous values of y, then $L(\lambda,\alpha,\theta,\sigma^2)$ is still the conditional log-likelihood of $y_1, \ldots, y_u$ given $x_1, x_0, x_{-1}, \ldots$ and $u_1, u_0, u_{-1}, \ldots$, and maximization of $L(\lambda, \alpha, \theta, \sigma^2)$ should still produce good estimates of $\lambda, \alpha, \theta, \sigma^2$. The distinction between conditional and unconditional estimates from time series data is discussed in Box and Jenkins (1976, Chapter 7). The dependence of $x_t$ and $u_t$ on past y may, however, produce biases. Walters (1985) found sizeable biases in a Monte Carlo study with small sample sizes, N = 10. His study assumed that $\lambda$ was known and equal to 0. His formulas show that the bias should decrease as N increases.

Following Box and Cox (1964), we maximize $L(\lambda, \alpha, \theta, \sigma^2)$ in two stages. First, for fixed $\lambda$ and $\alpha$ the MLE of $\theta$ is the nonlinear, weighted least-squares estimator $\hat{\theta}(\lambda, \alpha)$ which minimizes

$$SS(\lambda, \alpha, \theta) = \sum_{t=1}^{N} \{y_t^{(\lambda)} - f^{(\lambda)}(\underline{x}_t, \underline{\theta})\}^2 / u_t^{2\alpha}$$

and the MLE of $\sigma^2$ is $\sigma^2(\lambda, \alpha) = N^{-1} SS(\lambda, \alpha, \hat{\theta}(\lambda, \alpha))$. Define

$$L_{max}(\lambda, \alpha) = L(\lambda, \alpha, \hat{\theta}(\lambda, \alpha), \hat{\sigma}(\lambda, \alpha))$$

to be the log-likelihood maximized with respect to $\theta$ and $\sigma$ for fixed $\lambda$ and $\alpha$. An approximate MLE is found by computing $L_{max}(\lambda, \alpha)$ on a grid; in section 4 we use $\alpha = -1(.25)1$ and $\lambda = -1(.25)1$. If the exact MLE is needed then $L_{max}(\lambda, \alpha)$ can be maximized by a numerical optimization technique using the approximate MLE as an initial value. However, an exact MLE is probably unnecessary for most applications.

The function $L_{max}(\lambda, \alpha)$ can be used to test hypotheses and to construct confidence regions for $(\lambda, \alpha)$. For a given null value $(\lambda_0, \alpha_0)$, one can test $H_0$: $(\lambda, \alpha) = (\lambda_0, \alpha_0)$ against $H_1$: $(\lambda, \alpha) \neq (\lambda_0, \alpha_0)$ by the likelihood ratio test. The log-likelihood ratio is

$$LR(\lambda_0, \alpha_0) = L_{max}(\hat{\lambda}, \hat{\alpha}) - L_{max}(\lambda_0, \alpha_0).$$

Here $(\hat{\lambda}, \hat{\alpha})$ is the MLE or approximate MLE. $H_0$ is rejected at level $\epsilon$ if $2 LR(\lambda_0, \alpha_0)$ exceeds $\chi^2(1, 1-\epsilon)$, the $100(1-\epsilon)$ percentile of the chi-square distribution with one degree of freedom (Rao, 1973, section 6e). This test can be applied to each value $(\lambda_0, \alpha_0)$ on the grid, and a $100(1-\epsilon)$ percent confidence region for $(\lambda, \alpha)$ consists of all null values which are not rejected.

## 4. Examples

### 4.1 Population A

These data and the population B data discussed later were obtained through Professor Carl Walters (pers. comm.). Permission to identify the stocks has been refused by the original source. There are twenty-eight years of data. The variables, $R_t$ = total return and $S_t$ = spawner escapement, are plotted in figure 1.

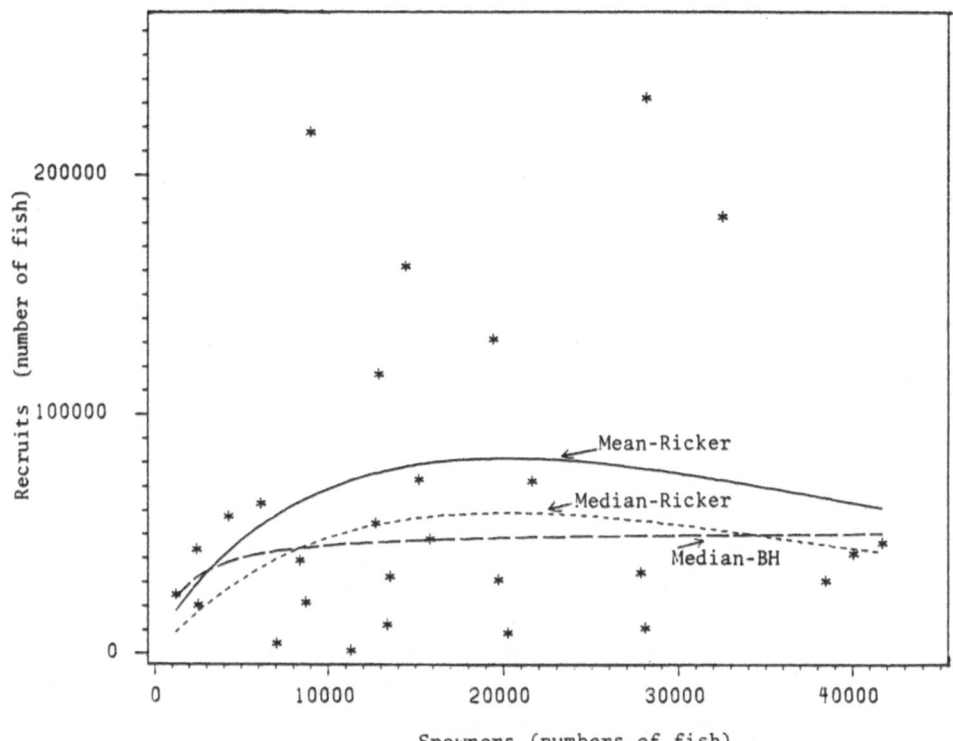

Fig. 1: Population A. Actual recruitment from 1940 to 1967,
estimated median recruitment based on the Ricker and
Beverton-Holt (BH) models, and estimated mean recruitment based on the Ricker model.

We first fit the (transformed) Ricker model:

$$R_t^{(\lambda)}/S_t^\alpha = (S_t \exp(\theta_1 + \theta_2 S_t))^{(\lambda)}/S_t^\alpha + e_t . \qquad (9)$$

The approximate MLE (on the grid $\lambda$ = -1(.25)1 and $\alpha$ = -1(.25)1) is $(\hat{\lambda}, \hat{\alpha})$ =
(.25,0) and the maximized log-likelihood is -338.3106. With $(\lambda, \alpha)$ equal to the
MLE, model (9) becomes

$$\sqrt[4]{R_t} = \sqrt[4]{S_t} \exp((\theta_1 + \theta_2 S_t)/4) + e_t .$$

Confidence regions for $(\lambda, \alpha)$ are given in Table 1. The 95% univariate confidence
regions for $\lambda$ and $\alpha$ are {0,.25,.5} and {-.25,0,.25,.5,.75}, respectively.

| Lambda | Alpha | | | | |
|---|---|---|---|---|---|
| | −.25 | 0 | .25 | .5 | .75 |
| 0 | * | * | * | | |
| | + | + | + | + | |
| .25 | * | ** | * | * | |
| | + | + | ++ | + | + |
| .5 | | * | * | * | * |
| | | + | + | + | + |

\* = 95% confidence region assuming the Ricker model.

\*\* = MLE assuming the Ricker model.

\+ = 95% confidence region assuming the Beverton-Holt model.

\+\+ = MLE assuming the Beverton-Holt model.

<u>Table 1</u>: Population A.  Confidence regions and maximum likelihood estimates of lambda and alpha.

Next the transformed Beverton-Holt model

$$R_t^{(\lambda)}/S_t^{\alpha} = \{1/(\theta_1 + \theta_2/S_t)\}^{(\lambda)}/S_t + e_t$$

was fit.  The MLE is $(\hat{\lambda}_1\ \hat{\alpha}) = (.25, .25)$ and the maximum log-likelihood is −337.0919.  The difference between the maximum log-likelihood for the Beverton-Holt and Ricker models is only 1.2187 which indicates that both models fit almost as well, though by this criterion the Beverton-Holt model does provide a slightly better fit.

The estimated median return, calculated as described in section 5, is plotted in figure 1 for both the Ricker and Beverton-Holt models, as is the "smearing estimate (section 5) of the mean assuming the Ricker model.

A researcher with only linear regression software might be tempted to linearize both the Ricker and Beverton-Holt models and then to compare them on their linearizing scales.  When the linearized Ricker model

$$\log(R_t/S_t) = \theta_1 + \theta_2 S_t$$

is, fit $R^2 = 0.207$ and the F-value for testing overall significance of the model is 6.79 (p = 0.015).  If the linearized Beverton-Holt model

$$1/R_t = \theta_1 + \theta_2/S_t$$

is fit then $R^2 = 0.000549$, F = 0.01, p = .9058, and if the alternative linearized model

$$S_t/R_t = \theta_2 + \theta_1 S_t$$

is fit then $R^2 = 0.110$, $F = 0.29$, and $p = .5950$. A researcher comparing these models only on their linearizing scales might easily be tempted into concluding that the Ricker model provides a far better fit. On the contrary, we believe that the Beverton-Holt model is very slightly better fitting, but the log transformation is vastly superior to the inverse transformation.

It is interesting to see how well the MLE transformation achieves both homoscedasticity and near normality. In table 2 the skewness and kurtosis are given for the three transformed Ricker models:

(I)      $R/S = \exp(\theta_1 + \theta_2 S)$   ($\lambda=1$, $\alpha=1$)

(II)      $\log(R) = \log(S) + \theta_1 + \theta_2 S$    ($\lambda=0$, $\alpha=0$)

(III)      $\sqrt[4]{R} = \sqrt[4]{S} \exp\{(\theta_1 + \theta_2 S)/4\}$     ($\lambda=.25$, $\alpha=0$),

that is, no power transformation, log transformation and the MLE, respectively.

| Lambda | Alpha | Skewness Correlation | Kurtosis | Spearman |
|--------|-------|-----------|----------|----------|
| 1 | 1 | .99 | 2.69 | .39 |
| 0 | 0 | −1.04 | 1.22 | .12 |
| .25 | 0 | −.27 | −.23 | .14 |

Table 2: Population A. Skewness and kurtosis of residuals and Spearman rank correlation between the absolute residuals and the predicted values. The Ricker model was used.

For model (III) both skewness and kurtosis are closest to 0, their theoretical value under normality. Also in table 2 are the Spearman rank correlations between the absolute residuals and the fitted values. Since the fitted values are an increasing function of S, the rank correlation is unchanged if the fitted values are replaced by S. Clearly, $\lambda=0$ and $\lambda=.25$ both transform to near homoscedasticity where there is little or no relationship between the mean and variance of R. However, the MLE $\lambda=.25$ is preferable to $\lambda=0$ since $\lambda=0$ "overtransforms" to negative skewness.

Normality and homoscedasticity of the residuals can be checked graphically by a normal plot of the residuals and a plot of the residuals against the fitted values. The graphical analysis of residuals should be done routinely for transformation models as for ordinary regression models. Draper and Smith (1981) provide an excellent account of graphical residual analysis. We plotted the residuals from the MLE and found no evidence of non-normality or heteroscedasticity. A normal probability plot of the residuals is roughly linear, though the very slight negative skewness (table 2) is evident. For these data, a plot of the residuals versus the predicted values is not easy to interpret; the Ricker and Beverton-Holt curves are nearly constant over the observed range of S so the predicted values are quite similar except for those corresponding to the smallest values of S.

## 4.2 Skeena River sockeye salmon

For this stock, effective escapement (S) and total return (R) are given for years 1940 to 1967 in Ricker and Smith (1975). These data are plotted in figure 2 along with the estimated medians (section 5) for the Ricker and Beverton-Holt models. Compared with the Population A stock, the Skeena River data show less positive skewness but more heteroscedasticity. Figure 2 shows several particularly low values of R associated with high values of S but only one particularly high value. This may indicate overcompensation, or it may simply be due to the high variability in R when S is large.

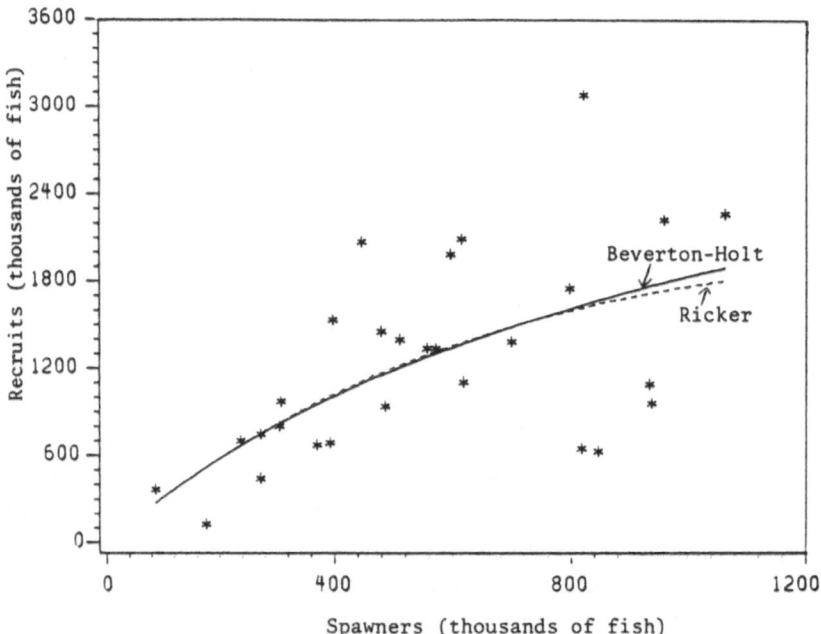

Fig. 2: Skeena River sockeye. Actual recruitment from 1940 to 1967 and estimated median recruitment based on Beverton-Holt and Ricker models.

The 95% confidence regions and the MLE of $(\lambda, \alpha)$ are the same for the Ricker and Beverton-Holt models and they are given in table 3. The maximum log-likelihoods for the Ricker and for the Beverton-Holt models differ by only 0.164, so there is little to suggest one model over the other. In particular, the data provide no strong evidence of overcompensation.

| Lambda | Alpha | | | | | |
|--------|-------|---|-----|----|-----|---|
|        | -.25  | 0 | .25 | .5 | .75 | 1 |
| 0      | *     | * |     |    |     |   |
| .25    | *     | * | *   | *  |     |   |
| .5     |       | * | *   | *  | *   |   |
| .75    |       | * | *   | ** | *   | * |
| 1      |       |   | *   | *  | *   | * |

* = 95% confidence region.

** = MLE

Table 3: Skeena River sockeye salmon stock. 95% confidence region and maximum likelihood estimate assuming either the Ricker or Beverton-Holt model.

Compared with the Population A data, the MLE here results in less transformation ($\hat{\lambda}$ = .75 instead of $\hat{\lambda}$ = .25 as before) but more weighting ($\hat{\alpha}$ = .5 instead of $\hat{\alpha}$ = 0 or .25). This is consistent with the observation that the untransformed Skeena data exhibit little skewness but considerable heteroscedasticity.

### 4.3 Pacific Cod, Necate Strait (Walters, et al., 1982)

The twenty-one observations, 1959 to 1979, on this stock are plotted in figure 3 along with predicted medians for two models.

Recruitment is virtually independent of spawning stock except that the two unusually large recruitments occur when S is rather small. For this reason the Beverton-Holt model fits poorly. For this stock we will consider the Ricker model and the power model

$$R_t = \theta_1 \, S_t^{\theta_2} .$$

With Box-Cox and weighting transformations the power model becomes

$$R_t^{(\lambda)}/S_t^{\alpha} = (\theta_1 \, S_t^{\theta_2})^{(\lambda)} / S_t^{\alpha} + e_t .$$

Confidence regions for ($\lambda$, $\alpha$) are given in table 4. The maximum log-likelihood is -34.9690 and -37.2541 for the power and Ricker models, respectively, and the large difference (2.2851) indicates lack-of-fit for the Ricker model.

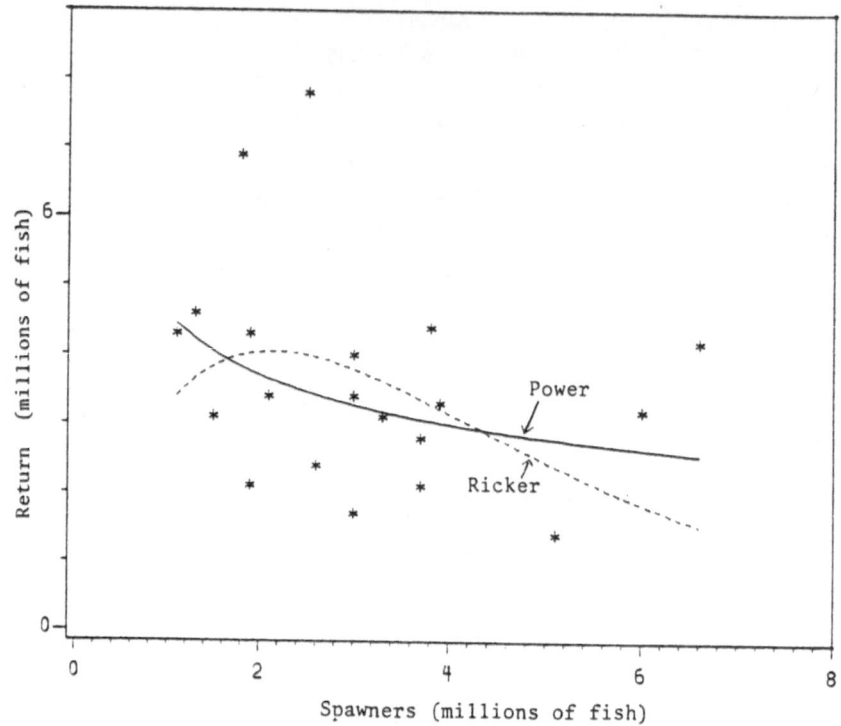

Fig. 3:  Hecate Strait Pacific Cod.  Actual recruitment
from 1959 to 1979 and estimated median recruitment based on the Power and Ricker models.

| Lambda | Alpha | | | | | | | | |
|---|---|---|---|---|---|---|---|---|---|
| | -1 | -.75 | -.5 | -.25 | 0 | .25 | .5 | .75 | 1 |
| -1 | | | | | | * | * | * | * |
| -.75 | | | | | * | * | * | * | * |
| -.5 | | | * | * | * | * | * | * | * |
| -.25 | | | * | * | * | * | * | * | * |
| 0 | | | * | * | * | * | ** | * | * |
| .25 | | | * | * | ** | ** | * | * | * |
| .5 | | * | * | * | * | * | * | * | |
| .75 | | * | * | * | * | * | * | | |
| 1 | | | * | * | * | * | | | |

* = 95% Confidence region.

** = MLE or within .1 of maximizing the log-likelihood.

Table 4.  Hecate Strait Pacific cod stock.  95% confidence region and maximum
likelihood estimate assuming the power model.

We can test the Ricker model as follows. The general model

$$R_t = S_t^{\theta_3} \exp(\theta_1 + \theta_2 S_t)$$

includes the Ricker model ($\theta_3 = 1$) and the power model ($\theta_2 = 0$) as special cases. One tests the Ricker model by testing $H_0$: $\theta_3 = 1$ against $H_1$: $\theta_3 \neq 1$. The maximum log-likelihood is $-34.9022$ and the log-likelihood maximized subject to the constraint $\theta_3 = 1$ is $-37.2541$ (see above). Twice the difference is $4.704$ which exceeds $X^2(1, .95) = 3.84$. The Ricker model is reject at level .05. By the same reasoning the power model is accepted since $34.9690 - 34.9022 = .0668$ is very small.

The 95% confidence region for ($\lambda$, $\alpha$) assuming the power model is quite large because (a) there are only 21 data points and, more importantly, (b) the variability in both spawning stock and return is small compared to the previously examined stocks. The confidence region for the Ricker model is even larger, but this is to be expected considering the lack of fit.

For the power model, the MLE is ($\hat{\lambda}$, $\hat{\alpha}$) = (0, .5), but the log-likelihood at $\lambda = .25$ and $\alpha = .25$ or .5 is within 0.1 of the maximum.

## 4.4 Population B (from Professor Carl Walters, pers. comm.)

For this stock, $R_t$ and $S_t$ vary over a much wider range than for the three preceding stocks, and it seems preferable to use the return to spawner ratio, $R_t/S_t$, as the response. A plot of $\log (R_t/S_t)$ against $S_t$ is rather linear, but most of the data are bunched together in the range $0 \leq S_t \leq 300,000$ while the remainder are scattered over $300,000 \leq S_t \leq 3,300,000$, so in figure 4 $\log (R/S)$ is plotted against $\log (S)$.

The only model studied here is the transformed Ricker model

$$(R_t/S_t)^{(\lambda)} / S_t^\alpha = \{\exp(\theta_1 + \theta_2 S_t)\}^{(\lambda)} / S_t^\alpha + e_t .$$

Also the grid $\alpha = -1(1/4)1$ is replaced by $\alpha = -1/4(1/16)1/4$ because values of $\alpha$ far from 0 fit poorly and lead to convergence problems when $\hat{\theta}_1$ and $\hat{\theta}_2$ are computed. The 95% confidence region for ($\lambda$, $\alpha$) and the MLE are found in table 5. Clearly, the confidence region is quite small. Because $R_t$ and $S_t$ vary over extremely wide ranges, $\lambda$ and $\alpha$ are well determined by the data.

## 5. Estimating the Conditional Distribution of y

We have seen how to utilize the model

$$y_i^{(\lambda)} / u_i^\alpha = f^{(\lambda)}(\underline{x}_i, \underline{\theta}) / u_i^\alpha + e_i$$

to efficiently estimate $\theta$. The model expresses a transformed response as a function of $\underline{\theta}$, $\underline{x}_i$, and the normally (or approximately normally) distributed $\epsilon_i$. Typically interest centers on the untransformed response $y_t$. In this section it is shown how to estimate the conditional (given $\underline{x}_i$) mean of $y_i$ as well as conditional quantiles such as the median.

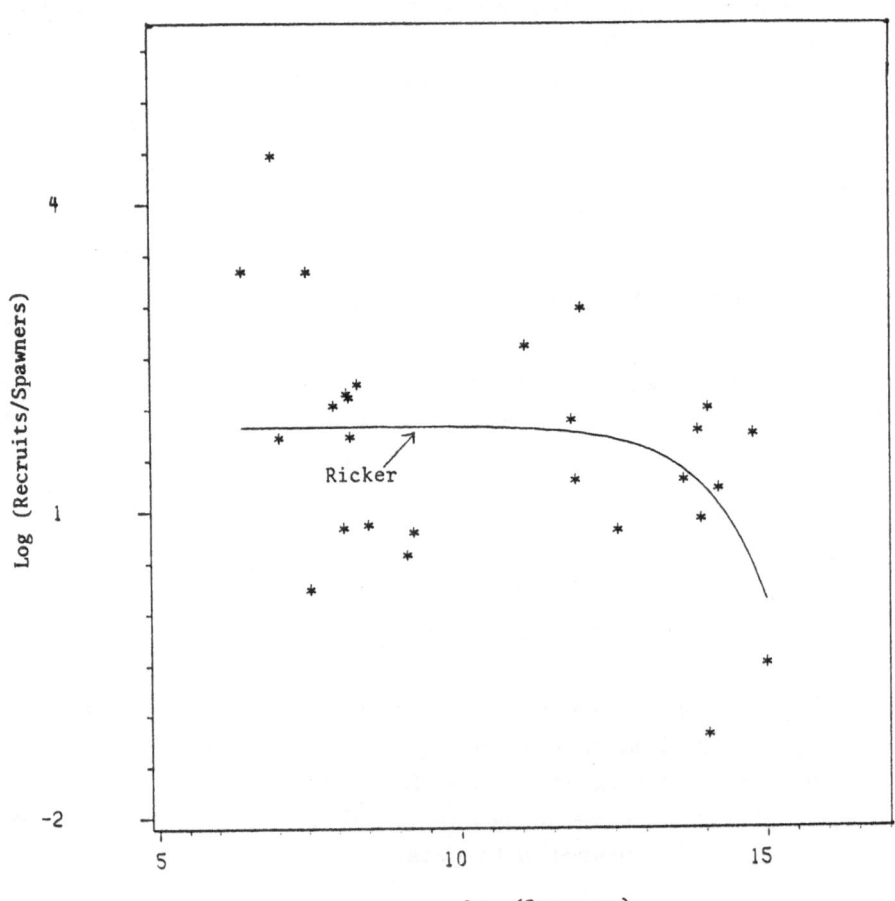

Fig. 4: Population B. Estimated median production ratio based on the Ricker model. Recruits and spawners are in numbers of fish. The production ratio and spawners are expressed in logarithms.

| Lambda | | | | Alpha | | | | |
|---|---|---|---|---|---|---|---|---|
| | -4/16 | -3/16 | -2/16 | -1/16 | 0 | 1/16 | 2/16 | 3/16 | 4/16 |
| -.25 | | * | * | | | | | | |
| 0 | | | | * | ** | * | * | |
| .25 | | | | | | * | * | * | * |

* = in 95% confidence regions

** = MLE

<u>Table</u> <u>5</u>. Population B. 95% confidence region assuming the Ricker model R/S = $\exp(\theta_1 + \theta_2 S)$.

Now $e_i$ cannot be exactly normally distributed in most cases, for example when $y_i$ and $f(x, \theta)$ are non-negative and $\lambda \neq 0$. It is better to suppose that $e_i$ is nearly normal but with a bounded range. Fortunately, for present purposes we need not make any assumption about the $\{e_i\}$ except that they are independent and identically distributed.

If $y^{(\lambda)} = x$, then the inverse relationship is $y = (1 + \lambda x)^{1/\lambda}$ if $\lambda \neq 0$ and $y = \exp(x)$ if $\lambda = 0$. In what follows we assume that $\lambda \neq 0$. If $\lambda = 0$, then one simply replaces $(1 + \lambda x)^{1/\lambda}$ by $\exp(x)$. Using (2) we can express the original response $y_i$ as a function of $\underline{x}_i$, $u_i$, $e_i$ and the parameters $\lambda, \alpha, \theta$:

$$y_i = \{1 + \lambda \, f^{(\lambda)}(\underline{x}_i, \underline{\theta}) + \lambda \, u_i^\alpha \, e_i\}^{1/\lambda} .$$

With this representation we can study the untransformed $y_i$. Let F be the distribution function of $e_1, \ldots, e_N$. The conditional mean of $y_i$ given $\underline{x}_i$ is

$$E(y_i|\underline{x}_i) = \int \{1 + \lambda \, f^{(\lambda)}(\underline{x}_i, \underline{\theta}) + \lambda \, u_i^\alpha \, e\}^{1/\lambda} \, dF(e) . \qquad (10)$$

Let $q_p$ be the pth quantile of F, i.e., $F(q_q) = p$. Then the conditional pth quantile of $y_i$ given $\underline{x}_i$ is

$$q_p(y_i|\underline{x}_i) = \{1 + \lambda \, f^{(\lambda)}(\underline{x}_i, \underline{\theta}) + \lambda \, u_i^\alpha \, q_p\}^{1/\lambda}$$

Duan (1983) has proposed the "smearing estimate" of $E(y_i|\underline{x}_i)$:

$$\hat{E}(y_i|\underline{x}_i) = N^{-1} \sum_{t=1}^{N} \{1 + \hat{\lambda} f^{(\hat{\lambda})} (\underline{x}_i, \underline{\hat{\theta}}) + \hat{\lambda} u_i^{\hat{\alpha}} \hat{e}_t\}^{1/\hat{\lambda}} \tag{11}$$

—where $e_t$ is the t-th residual,

$$\hat{e}_t = \{y_t^{(\hat{\lambda})} - f^{(\hat{\lambda})}(\underline{x}_t, \underline{\hat{\theta}})\} / u_t^{\hat{\alpha}} .$$

The relationship between (î0) and (11) is clear; all parameters are replaced by their estimates and averaging with respect to the theoretical distribution of $e_i$ is replacing by averaging over the sample $\hat{e}_1, \ldots, \hat{e}_N$. Even if F were known, (9) could only be evaluated by numerical integration, but (10) does not require integration. This is a distinct advantage of the smearing estimate.

Let $\hat{q}_p$ be the p-th sample quantile of $\{\hat{e}_i\}$. Then we estimate $q_p(y_i|\underline{x}_i)$ by

$$\hat{q}_p(y_i|\underline{x}_i) = \{1 + \hat{\lambda} f^{(\hat{\lambda})}(\underline{x}_i, \underline{\theta}) + \hat{\lambda} u_i^{\hat{\alpha}} \hat{q}_p\}^{1/\hat{\lambda}} .$$

In the case of the median (p = .5), the residuals should have a median close to 0 and we can replace $\hat{q}_{.5}$ by 0. The estimated conditional median is then

$$\hat{m}(y_i, \underline{x}_i) = \hat{q}_{.5}(y_i|\underline{x}_i)$$

$$= \{1 + \hat{\lambda} f^{(\hat{\lambda})}(\underline{x}_i, \underline{\theta})\}^{1/\hat{\lambda}} = f(\underline{x}_i, \underline{\hat{\theta}}) .$$

As mentioned before, in figure 1 $\hat{E}(y|\underline{x}_i)$ and $\hat{m}(y|\underline{x})$ are plotted for Population A assuming the Ricker model, and $\hat{m}(y|x)$ is plotted for the Beverton-Holt model as well. The mean return is always considerably larger than the median return. This reflects the considerable positive skewness seen in the actual recruitments and evident in the MLE of $\lambda$, $\hat{\lambda} = .25$.

Because recruitment is so highly variable any realistic management model will be stochastic, and when a stochastic model is constructed it is vital that the entire conditional distribution of $R_t$ given $S_t$ be estimated. This can be done by estimating conditional quantiles as above. Ruppert, Reish, Deriso, and Carroll (1985) use a closely related method for estimating conditional quantiles when constructing a stochastic model of the Atlantic menhaden population. Transformation of the menhaden stock-recruitment data is discussed further in Carroll and Ruppert (1984).

## Summary and Conclusions

A theoretical model relating a response y to independent variables $\underline{x}$ and parameters $\underline{\theta}$ may not be suitable in its original form for least-squares estimation. This is the case if the response exhibits skewness or nonconstant variance.

However, if the response and the model function are transformed in the same way, then the transformed response may be approximately normally distributed with a nearly constant variance and then least-squares estimation will be efficient.

In this paper we propose combining the Box-Cox power transformation with weighting by $u_t^{\alpha}$, where $u_t$ is a variable not depending on $y_t$ and $\alpha$ is a parameter to be estimated. Another possibility, one that is not explored here, is to have $u_t$ be the predicted value $f(\underline{x}_i, \hat{\underline{\theta}})$. Weighting the untransformed response by $|f(\underline{x}_i, \hat{\theta})|^{\alpha}$ is studied in Box and Hill (1974), Pritchard, Downie, and Bacon (1977) and Carroll and Ruppert (1982).

The Box-Cox parameter $\lambda$ and the power weighting parameter $\alpha$ can be estimated simultaneously with $\underline{\theta}$ and $\sigma$ by maximum likelihood. A confidence region for $(\lambda, \alpha)$ can be constructed by likelihood-ratio testing.

Four stock-recruitment data sets were analyzed by our transformation methodology. The original response, return, is in each case skewed or heteroscedastic. Only for the Pacific cod stock is $\lambda=1$ and $\alpha=0$, corresponding to no transformation, in the 95% confidence region. On the other hand $\lambda=0$ and $\alpha=0$, which is the linearizing transformation for the Ricker model, is in the 95% confidence region for all four stocks. Also, the inverse transformation $(\lambda = -1)$ which linearizes the Beverton-Holt is not in the 95% region for any of the four stocks. If both the Ricker and Beverton-Holt models are linearized, then the Ricker model will appear better fitting, not because it is necessarily superior but because the log transformation is more suitable than the inverse transformation. There are examples where $\lambda=-1$ is quite suitable, for example the Atlantic menhaden stock (Carroll and Ruppert, 1984).

After $\lambda$ and $\alpha$ have been estimated by maximum likelihood, the fitted model should be checked by residual analysis as described in, for example, Draper and Smith (1981). Maximum likelihood estimatiion is highly sensitive to outlying observations. Such influential points may be evident from the residuals. Robust estimators for our transformation model is an important area for future research. At present, robust estimation has been studied only for the rather different methodology where only the response, not the model, is transformed; see Carroll and Ruppert (1985), Carroll (1980), and Bickel and Doksum (1981).

The model function gives the conditional median of the response, and the conditional mean and the other conditional quantiles of the response can be easily estimated. By estimating conditional quantiles one in effect builds a model of the skewness and heteroscedasticity in the untransformed response. Such a model is a crucial part of a realistic stochastic management model of the stock.

We have used four fish stocks as examples of our proposed statistical methodology, but our analyses should not be considered definitive. For example, we did not consider the effects on the Skenna River stock of the 1951-2 rock slide, changes in exploitation rates, an artifical spawning channel opened in 1965, or interactions between year classes; see Ricker and Smith (1975). We believe, however, that power and weighting transformations will be equally as useful for more elaborate models as for the basic ones that we have employed for illustrative purposes.

Acknowledgments: Rod Reish and Rick Deriso introduced us to stock-recruitment analysis. Carl Walters supplied the data, and we benefitted from his helpful discussions. The research of R.J. Carroll was supported by Air Force Office of Scientific Research contract AFOSR F49620-82-C-0009. The work of D. Ruppert was supported by National Science Foundation Grant MCS-87-3129.

REFERENCES

Bartlett, M.S., 1947. The use of transformations. _Biometrics_, 3, 39-52.

Beverton, R.J.H., and Holt, S.J., 1957. _On the dynamics of export fish populations._ Her Majesty's Stationery Office, London, 533 p.

Bickel, P.J., and Doksum, K.A., 1981. An analysis of transformations revisited _Journal of the American Statistical Association_, 76 296-311.

Box, George E.P. and Cox, David R., 1964. An analysis of transformations (with discussion). _Journal of the Royal Statistical Society, Series B_, _26_, 211-246.

Box, George E.P. and Hill, William J., 1974. Correcting inhomogeneity of variance with power transformation weighting. _Technometrics, 16_, 385-389.

Box, George E.P. and Jenkins, G.M., 1976. _Time Series Analysis: Forecasting and Control, Revised Edition._ Holden-Day: San Francisco.

Carroll, R.J., 1980. A robust method for testing transformations to achieve approximate normality. _Journal of the Royal Statistical Society, Series B 42_, 71-78.

Carroll, R.J. and Ruppert, D., 1982. Robust estimation in heteroscedastic linear models. _Annals of Statistics 10_, 429-441.

Carroll, R.J. and Ruppert, D., 1984. Power transformations when fitting theoretical models to data. _Journal of the American Statistical Association_, 79, 321-328.

Carroll, R.J. and Ruppert, D., 1985. Transformations in regression: a robust analysis. _Technometrics_, 27, 1-12.

Draper, Norman and Smith, Harry, 1981. _Applied Regression Analysis, 2nd Edition._ Wiley: New York.

Duan, Naihua, 1983. Smearing estimate: a nonparametric retransformation method. _Journal of the American Statistical Association_, 78, 605-610.

Pritchard, D.J., Downie, J., and Bacon, D.W., 1977. Further consideration of heteroscedasticity in fitting kinetic models. _Technometrics_, 19, 227-236.

Ricker, W.E., 1954. Stock and recruitment. J. of Fish. Res. Board Can., 11, 559-623.

Ricker, W.E. and Smith, H.D., 1975. A revised interpretation of the history of the Speena River sockeye salmon (Oncorhynchus nerka). J. the Fish. Res. Board of Can., 32, 1369-1381.

Ruppert, D., Reish, R.L., Deriso, R.B., and Carroll, R.J., 1985. A Stochastic population model for managing the atlantic manhaden fishery and assessing managerial risks. Can. J. Fish. and Aqu. Sc. - to appear.

Walters, C.F., 1985. Bias in the estimation of functional relationships from time series data. Can. J. Fish. and Aqu. Sc., 42, 147-149.

Walters, C.F., Hilborn, R., Staley, M.J., Wong, F., 1982 Report to the Pearse Commission on Pacific Fisheries Policy, Vancouver, B.C.

# A CONCEPTUAL MODEL FOR MULTISPECIES, MULTIFLEET FISHERIES

Wayne M. Getz
Division of Biological Control
University of California, Berkeley, CA 94720

Gordon L. Swartzman
Center for Quantitative Sciences in Forestry,
    Fisheries and Wildlife Management
University of Washington
Seattle, WA 98112

Robert C. Francis
Northwest and Alaska Fisheries Center
NMFS
7600 Sand Point Way NE
Seattle, WA 98115

## Introduction

Marine fisheries scientists and managers are becoming more cognizant of the need to regard the exploitation of many single species stocks as part of a larger multispecies system (see, for example, Mercer 1982 and the 26 papers published therein).  A problem in the analysis and management of any ecological and/or economic system is demarcating its boundary: there will always be components external to the bounded system that strongly influence its behavior and yet are dealt with in a simple manner often as a time varying input function.  The approach that needs to be taken, especially with regard to managment problems, is to be pragmatic, to strike a balance between complexity and utility and to be goal oriented.  If the goal is to understand the biological mechanisms that determine larval survival rates and hence recruitment levels, then comprehensive ecologically based models are required (cf. Laevastu, Favorite and Larkins 1982).  If the primary goal is to assess stock and yield levels from catch and effort data, then a less biologically detailed approach is appropriate (cf. Murawski 1984, Ursin 1982).

In this paper, a conceptual framework is proposed for modeling the behavior of a multispecies, multifleet interaction in which the interspecies dynamics are linked by fleet behavior rather than strongly identifiable ecological interactions.  The latter has been examined using the Antarctic krill/marine mammal fishery (May et al. 1979) as its exemplar.  Our exemplar will be the ground-fish fishery that operates off the Oregon and Washington coasts of the western United States.  This fishery includes a number of rock fish and flat fish species, the valuable long-lived sable fish and interacts strongly at the fleet level with local shrimp fisheries.  Although this complex is a guide in conceptualizing the model presented below, the approach taken is otherwise general.  Its orientation, however, is rather specific. The goal is to develop a framework that can be used to address particular questions relating to the regulation of multifleet multi-species fisheries. Thus, the framework includes structures, such as age, which interact strongly with fishing technologies and have important marketing implications. It also relies on parameters that can be estimated from boat log books and trip-ticket data, where available. Market forces that determine price and costs are regarded as external to the framework, i.e., price and costs are regarded as inputs. Internalization of market dynamics requires embedding the fleet/species dynamics in a socio--economic setting that encompasses other fisheries, consumer population structures, etc. This level of complexity is avoided so that the fleet dynamics component can be focused upon.

Population dynamics

Several questions need to be addressed in proposing an appropriate framework for modeling the multispecies population dynamics component of the fishery.

Firstly, should lumped or age-structured population models be used? Lumped population models, such as Ricker's surplus production model, have played an important role in developing a bioeconomic theory in fisheries management (Clark 1976, Clark and Munro 1980, McKelvey 1982a, 1982b). Age structure, however, plays an important role in determining catch levels as a function of different technologies, and the value of that catch to the processing industry. Thus, for fisheries where catch-at-age data is available, if the time series is adequate to support an age-structured model then age structure should be included. In some cases, even if the data is inadequate, age-structured models, in which age specific mortality parameters are set constant across age, can prove invaluable for exploring age-related questions such as juvenation as an index of overexploitation in the fishery. Under certain assumptions age-structured models can be collapsed into paired difference equations from which the age structure can be reconstituted (see Deriso 1980 for details).

Secondly, should ecological interactions be included? Large scale ecosystem models have been developed to assess the productivity of the fishery resources in the Northeastern Pacific including the North American West Coast groundfish fisheries. These models however are too complex to embed in a multifleet economic study of a fishery. Further, unless ecologically significant interactions can be identified and realistically structured in the multispecies model (such as a significant predation interaction between species) a basic Beverton and Holt type age structure model will suffice. The latter is justified by the overall lack of precision in the process of estimating model parameters such as age-specific natural mortality rates and annual recruitment levels from catch-at-age versus effort data, as a function of environmental conditions using techniques such as VPA analysis (Pope 1972).

Consideration of these questions suggest that the following approach to modeling species dynamics is appropriate. Let $N_{ij}(t)$ be the number of individuals in age class $i$ of species $j$. Let $n_j$ be the number of age classes in species $j$ and $m_s$ be the total number of species (or ecologically separated stocks) in the fishery.

We will assume Beverton and Holt (1957) fisheries dynamics, i.e.,

$$\frac{dN_{ij}}{dt} = -(M_{ij}(t)+F_{ij}(t))N_{ij} \qquad i=1,...,n_j \qquad j=1,...,m_s, \qquad [1]$$

where $M_{ij}(t)$ and $F_{ij}(t)$ respectively are the natural and fishing mortality rates on the population at time $t$. In our groundfishery exemplar, catch/effort data and the analyses of fleet behavior and stock availability have been based on quarterly (seasonal) time periods. Thus, if the units of $t$ are years, we will choose $t=1/4$ as the basic time iteration unit. The analysis presented below, however, is valid for any reasonable subdivision of the year including an annual iteration interval if seasonal data is unavailable. Integrating [1], assuming the $M_{ij}$'s are seasonal (i.e., may vary from season to season but not year to year) and that $M_{ij}$ and $F_{ij}$ are constant over a quarter but are expressed in terms of an annualized rate, we obtain

$$N_{ij}(t+(\ell+1)/4) = N_{ij}(t+\ell/4)e^{-[M_{ij}(\ell/4)+F_{ij}(t+\ell/4)]/4}$$
$$i=1,...,n_j, \qquad j=1,..., m_s, \qquad \ell =0,1,2 \quad \text{and}$$
$$t = 0,1,2,... \qquad [2]$$

When $\ell= 3$ aging takes place. Thus, for $j=1,...,m_s$, $t=0,1,2$,

$$N_{i+1j}(t+1) = N_{ij}(t+3/4)e^{-[M_{ij}(3/4) + F_{ij}(t+3/4)]/4}$$
$$i=1,..., n_j-2 \qquad [3]$$

and

$$N_{n_jj}(t+1) = N_{n_jj}(t+3/4)e^{-[M_{n_jj}(3/4) + F_{n_jj}(t+3/4)]/4}$$
$$+ N_{n_j-1j}(t+3/4)e^{-[M_{n_j-1j}(3/4)+F_{n_j-1j}(t+3/4)]/4} \qquad [4]$$

where $N_{n_jj}$ is the number of individuals of age $n_j$ and older. Note throughout the text that $t$ will be used depending on the context, to denote both continuous time $t>0$ and discrete time $t=0,1,2,...$ .

Assume that recruitment to the youngest age class (not necessarily to the fishery, depending on catchability) is best correlated with fecund stock biomass levels at the beginning of a particular one of the four preceding seasons (if this is inappropriate a time lag must be introduced). Let $w_{ij}(t)$ be the average weight of an individual in age class $i$ of species $j$ and let $c_{ij} > 0$ be its average fecundity (taking sex ratio into account). Then fecund stock biomass $S_j(t)$ is given by

$$S_j(t) = \sum_{i=1}^{n_j} c_{ij}w_{ij}(t) N_{ij}(t) \qquad [5]$$

Using stock-recruitment data, often generated from a VPA analysis of the raw catch-effort data (Pope 1972), it is possible to fit a stock recruitment relationship, viz.

$$N_{1j}(t+1) = f_j(S_j(t+\ell_j/4)) \quad j=1,...,m_s \qquad\qquad [6]$$

where $\ell_j = 0,1,2$ or $3$ (whichever is most appropriate and $f_j$ has some suitable a apriori selected form (e.g., Beverton and Holt, Ricker or depensatory form—cf. Getz (1980a)). In general a deterministic relationship, such as [6], provides a poor fit to the data and a stochastic approach to modeling recruitment in terms of stock is more appropriate (cf. Getz 1984). This makes the analysis much more complex. Thus stochastic considerations will be ignored so as not to detract from the expository nature of the material presented here.

Once the stock recruitment relationships $f_j$ and natural mortality coefficients $M_{ij}(\ell/4)$ have been estimated the model defined by equations [1]-[6] can be used to simulate the impact of fishing mortalities $F_{ij}(t+\ell/4)$ $\ell = 0,1,2,3$, $t=0,1,2,...$ on the populations starting from known initial conditions $N_{ij}(0)$ where $i=1,...,n_j$ and $j=1,...,m_s$. Notice that the j species are dynamically separate in the absence of fishing (i.e., $F_{ij} = 0$ for all i and j). Thus, species interactions depend on the links that exist between the various fishing mortality rates $F_{ij}$ as generated at one level by the particular fishing technologies employed (hard links imposed by catchability constraints), and at a second level by prices (soft links generated by market conditions).

## Fleet Characterization

A major challenge in multifleet fisheries is to model the diversity of the fleet in terms of its response to the processing industry (market conditions) and its impact on the stock while, at the same time, retaining a simplicity that facilitates comprehension of the fleet/stock interaction. To achieve this we assign boats to one of $m_f$ classes, where it is assumed that all boats in technology class k employ the same fishing technologies (i.e., are subject to the same set of catchability coefficients) but may induce different fishing mortality rates per boat per day due to vessel size, skipper skill or efficiency at locating fish. In each class k , $k=1,...,m_f$ , boat r , $r=1,...,b_k$ will at any time t be involved in one of activities $A_\nu$ . In general the range of $\nu$ is determined by the problem on hand. In the treatment below it will be assumed that there are $m_s+2$ categories associated with each boat, i.e., $\nu = 0,...,m_s+1$, as follows. $A_0$ is an idle boat (possibly undergoing repairs), $A_\nu$ $\nu = 1,..,$ $m_s$ is a boat targeting on species $\nu$ and $A_{m_s+1}$ is a boat participating in its most profitable option external to the multispecies fishery as bounded by species 1 to $m_s$ . Note that although $A_0,...,$ $A_{m_s}$ is common to all boats . $A_{m_s+1}$ may be particular to each boat in which case we have a set of activities $A_{rkm_s+1}$ for $k=1,...,m_f$ and $r=1,...,b_k$ .

Following our assumptions in setting up the boat class structure, all boats in class k targeting on species $\nu$ experiences catchability coefficients $q_{ij}(t)$ at time t. We will assume that the value of these coefficients depends only on the season and is otherwise independent of t. Although this still implies that a large number of catchability coefficients may need to be estimated, it follows that $q_{ijk\nu} = 0$ for all $j \neq \nu$ $j=1,...,$ $m_s$ unless boats in class k experience by-catch problems when targeting on a particular species. Thus, the number of catchability coefficients really depends on the number of different fishing technologies employed, the extent to which by-catch is experienced while targeting, and on the season under consideration. An analysis of species composition in landings does not directly indicate by-catch events since a particular boat may target on more than one species during one trip.

The performance of a boat can be divided into two components. Firstly, its ability to translate fishing days of a particular activity into fish mortality and hence calculate catch and gross revenues. This ability will be referred to as the boat's "power". Secondly, the cost associated with participating in a particular activity, hence allowing fishing days for a set of particular activities to be finally translated into net revenues. Associate a record $R_{rk}$ with the r-th boat in the k-th technology class at time $t$ that specifies a power vector $\underline{\rho}_{rk}$ of elements $\rho_{rk\nu}$ $\nu=1,...,m_s$, and for each $\nu=0,...,m_s+1$ a set of cost parameters $c_{rk\nu}$, i.e.,

$$R_{rk} = \{\underline{\rho}_{rk}{}^{(t)}, c_{rk\nu} \quad \nu=0,...,m_s \} \quad r=1,...,b_k, \quad k=1,...,m_f$$

More generally, $R_{rk}$ may also contain information relating to age, repair record and other historical information that could be used to modify cost parameter values etc. The role of parameters $c_{rk\nu}$ will be discussed more fully in a later section.

The parameters $\rho_{rk\nu}$ are estimated from catch data as a function of both stock abundance and boat days for boats in the same class performing the same activity on stocks. Hence, if boat rk (henceforth rk will be used to denote boat r in class k ) is targeting on species $\nu$ then, assuming as in Beverton and Holt (1957) that catch rate is proportional to stock density, it follows that individuals in the i-th age class of species j are caught at a rate $K_{ijrk\nu}(t)$ given by

$$K_{ijrk\nu} = \rho_{rk\nu} q_{ijk\nu}(t) N_{ij}(t) .$$ [7]

The units of $\rho_{rk\nu}$ are chosen so that the units of [7] are annualized catch rate. Thus, at time $t$, the fishing mortality rate in equation [1] induced by the activity of all boats at time $t$ is

$$F_{ij\nu}(t) = \sum_{k=1}^{m_f} \sum_{r=1}^{b_k} \rho_{rk\nu} q_{ijk\nu}(t) \qquad j=1,...,m_s, \quad i=1,...,n_j$$ [8]

where $\nu = \nu_{rk}(t)$ is the particular activity chosen by boat rk at time $t$. If we assume during a particular quarter $\ell$, $\ell=0,...,3$ that boat rk spends an annualized proportion $(d_{rk\nu})$ of days on activity $\nu=1,...,m_s$, and that catchability coefficients are constant over the season then, on average, the annualized fishing mortality for the $\ell$-th quarter in year $t$ is

$$F_{ij}(t+\ell/4) = \sum_{k=1}^{m_f} \sum_{r=1}^{b_k} \sum_{\nu=1}^{m_s} d_{rk\nu}(t+\ell/4) \rho_{rk} q_{ijk\nu}(\ell/4)$$ [9]

$$\ell = 0,1,2,3, \quad j=1,...,m_s, \quad i=1,...,n_j$$

Similarly from equation [7] the number of individuals in the i-th age class of the j-th species caught by boat rk during the $\ell$-th quarter of the t-th year is

$$K_{ijrk}(t+(\ell+1)/4) = \sum_{\nu=1}^{m_s} d_{rk\nu}(t+\ell/4)\, \rho_{rk\nu}q_{ijrk}(\ell/4)\, N_{ij}(t)dt \qquad [10]$$

which, from 1 it follows that

$$K_{ijrk}(t+(\ell+1)/4) = \sum_{\nu=1}^{m_s} d_{rk\nu}(t+\ell/4)\, \rho_{rk\nu}q_{ijk\nu}(\ell/4)N_{ij}(t+\ell/4)$$

$$(1-e^{-[M_{ij}(\ell/4)\,+\,F_{ij}(t+\ell/4)]/4})/M_{ij}(\ell/4)\,+\,F_{ij}(t+\ell/4) \qquad [11]$$

The total biomass yield obtained from the j-th species for the $\ell$-th season in year t is the sum of the number of individuals caught in each age class during that period multiplied by the average weight for that age class ($w_{ij}(\ell)$). Thus the biomass yield of species j taken by boat rk for the $\ell$-th season in year t is

$$Y_{jrk}(t+(\ell+1)/4) = \sum_{i=1}^{n_j} w_{ij}(\ell)K_{ijrk}(t+(\ell+1)/4) \qquad \ell=0,1,2,3,$$
$$j=1,...,m_s,\ k=1,...,m_s,\ r=1,...,b_k \qquad [12]$$

By summing [12] over one or more of the indices $\ell$, j, k, r, one can calculate quarterly yield totals across species, across boats in a given technology class, across technology classes, as well as annual values for each of these cases.

## Catchability, efficiency and sustainable yield

If, in a deterministic model of the form [2]-[6], the fishing mortality rates depend only on $\ell=0,1,2,3$ but not on t, i.e., for a particular season mortality rates are constant from year to year, then for a set of fishing mortality rates $F_{ij}(\ell/4)$, $\ell=0,1,2,3$, $j=1,...,m_s$ and $i=1,...,n_j$ it may be possible to find solutions for each j satisfying

$$N_{ij}(t+1+\ell) = N_{ij}(t+\ell)\ i=1,...,n_j\ ,\ \ell = 0,1,2,3 \qquad [13]$$

may exist. These solutions are called equilibrium solutions and their number and stability properties are essentially dependent on the forms of the stock-recruitment relationships $f_j$ appearing in equation [6] (Levin and Goodyear 1980, Reed 1980, Getz 1980a). If these $f_j$'s are compensating then the equilibrium soluitions are unique and stable (Levin and Goodyear 1980) and correspond to a sustainable biomass yield $Y_j(\{F_{ij}(\ell/4) \mid i=1,...,n_j,\ \ell=0,1,2,3\})$ that either has a maximum $Y_j^o$ for an optimal choice of $F_{ij}(\ell/4) = F_{ij}^o(\ell/4)\ i=1,...,n_j, \ell=0,1,2,3$ or is bounded above by and approaches $Y_j^o$ asymptotically as one or more of the $F_{ij}$'s approach $\infty$ in a particular way (Getz 1979). A maximum sustainable biomass yield $Y_j^o$

for species $j$ often exists for depensatory and overcompensatory stock–recruitment relationships $f_j$ as well. Its realization depends on whether it is possible to independently manipulate the fishing mortalities $F_{ij}(\ell/4)$. For a single species $j$, the mortalities rates $F_{ij}$ are usually linked with respect to $i$, since fishing technologies invariably are unable to separate completely the age classes in the catch. In addition, in multispecies fisheries the $F_{ij}(\ell/4)$'s will be linked across $j$ if by–catches occur. Further the values of $F_{ij}$ are constrained by the number and types of boats in the fishery. Thus, in general, sustainable yields $Y_j^0$ cannot be achieved and for this reason, these yields have been referred to as ultimate sustainable yields (USY) (Getz 1980b). Thus we will refer to $\{Y_j^0 \mid j=1,...,m_s\}$ as the set of USY's for the multispecies fishery under consideration.

In terms of equation [9] and equilibrium harvesting, if we define the total fishing effort in quarter $\ell$ of all boats in class $k$ targeting on species $\nu$ as

$$e_{k\nu}(\ell/4) = \sum_{r=1}^{b_k} d_{rk\nu}(\ell/4)\, \rho_{rk\nu} \qquad\qquad [14]$$

then

$$F_{ij}(\ell/4) = \sum_{k=1}^{m_f} \sum_{\nu=1}^{m_s} q_{ijk}(\ell/4)\, e_{k\nu}(\ell/4) \qquad \begin{array}{l} \ell=0,1,2,3,\ j=1,...,m_s \\ i=1,...,n_j \end{array} \qquad [15]$$

Thus our ability to manipulate the $F_{ij}$'s depends on the relative values of the catchability coefficients with respect to technology and targeting parameters $k$ and $\nu$ within and between seasons $\ell$ and is also limited by the constraints $e_{k\nu} > 0$ for all $k$ and $\nu$. If we assume that j-th species does not occur as a by–catch with any other species then only $q_{ijkj} \neq 0$ and [15] can be written in matrix form as

$$\underline{F}_j(\ell/4) = Q_j(\ell/4)\, \underline{e}_j(\ell/4) \qquad \ell=0,1,2,3 \qquad\qquad [16]$$

where $\underline{F}_j = (F_{ij},...,F_{n_jj})^T$, $\underline{e}_j = (e_{ij}...e_{m_sj})^T$ and $Q_j$ is a matrix of non-negative elements $(q_{ijkj})$. From matrix theory, the $F_{ij}(\ell/4)$'s can only be uniquely determined for a suitable choice of $e_{kj}(\ell/4)$'s if the rank of $Q_j(\ell/4)$ is $n_j$, that is, at least $n_j$ technologies exist (i.e., $m_s \geq n_j$) and their catchability coefficients are linear independent. If these two conditions are satisfied the mix of technologies will be referred to as complete for the exploitation of the j-th species in the $\ell$-th season. However, a mix of technologies that are complete for species $j_1$, may not be complete for species $j_2$ and by–catch problems may further complicate the situation.

Consider the yield from the j-th stock, where the $F_{ij}$'s are determined by the choice of effort levels $e_{kj}(\ell/4)$ in [15], $\ell=0,1,2,3$, $k=1,...,m_f$ that maximize $Y_j(\{F_{ij}(\ell/4)\ i=1,...n_j, \ell=0,1,2,3\})$. If $Y_j^*$ and the $e^*_{kj}(\ell/4)$'s denote these choices then it follows that $Y_j^* < Y_j^0$ since the $F_{ij}$'s are linked across $i$ (i.e., by age class). Suppose the mix of technologies is incomplete so that, in general, $Y_j^* < Y_j^0$. Define the biomass exploitation efficiency $E_j$ for a particular mix of technologies exploiting species $j$ as

$$E_j = Y_j^* / Y_j^0 \quad j=1,...,m_s \tag{17}$$

Then it is still possible for single technologies alone to perform well in terms of obtaining an $E_j > 0.9$ (cf. Getz 1980b). In a multispecies fishery, if by-catches are prevalent, then the yield from each species cannot be independently considered and an overall efficiency can be considered in terms of

$$E = (\sum_{j=1}^{m_s} Y_j)^* / \sum_{j=1}^{m_s} Y_j^0 \tag{18}$$

where

$$Y^* = (\sum_{j=1}^{m_s} Y_j)^* \tag{19}$$

is obtained by the choice of $e_{k\nu}(\ell/4)$ in [15] that maximizes the sum of the yields from the individual species. Note, in general, that $(\sum_{j=1}^{m_s} Y_j)^* < \sum_{j=1}^{m_s} Y_j^*$. It is also worth noting that indices $E_j$ and $E$ respectively in [17] and [18] do not directly pertain to questions related to the economics of exploiting species within the fishery.

## Bioeconomics of the fleet

The economics of operating a boat in a multifleet multispecies fishery is exceedingly complex. In the previous sections we developed a multispecies population dynamics model and identified the $F_{ij}$'s in equation [15] as the interface between fleet effort levels and actual catch rates (cf. equations [9] and [11]). To provide a framework for modeling dynamics at the next hierarchical level, i.e., the response of the fleets to the biological state of the fishery and to the economic inputs, several simplifying assumptions need to be made. Firstly, we will assume that the number of days that a boat participates in a particular activity can be averaged over a quarter. We will also assume that the economic parameters involved with operating costs and prices that apply during a quarter are known to individual boat or fleet operators at the beginning of that quarter.

As discussed previously, we assume that short term economics considerations, in operating a boat, depend on its cost and revenue streams.

Consider a particular activity $\nu$. In defining record $R_{rk}$ we indicated that for boat rk there are cost parameters $c_{rk\nu}$ associated with its participation in this activity during quarter $\ell$.

In the following analysis we will omit arguments associated with time and focus on the economics related to operating a boat over a given quarter. Essentially we will reduce the fleet dynamics problem to the solution of one or more static non-linear mathematical programming problems. The formulation can then be extended to a dynamic decision making setting to address problems defined over a longer time horizon.

First assume that it is possible to assign or to estimate from data a cost relationship (regression) for each boat with respect to each activity as a function of the number of days spent in that activity during a quarter. Thus for the boat rk, if the cost $C_{rk\nu}$ is a linear function of days $d_{rk\nu}$ per quarter spent on activity $\nu$, it will have the form

$$C_{rk\nu}(d_{rk\nu}) = c^1_{rk\nu} + c^2_{rk\nu}d_{rk\nu} \qquad \nu=0,\ldots,m_s+1 \ , \qquad\qquad [20]$$

where both $c^1_{rk\nu}$ and $c^2_{rk\nu}$ are assumed to be non-negative. Note that $d_{rk0}$ and $d_{rkm_s+1}$ respectively are the number of days spent idling in port and on the most profitable activity external to the fishery. The two constants in [20] respectively can be interpreted as the fixed and variable costs associated with activity $\nu$ . In general, however, $C_{rk\nu}(d_{rk\nu})$ may be nonlinear, e.g., reducing costs to scale.

Similarly, a profit function can be constructed, but one important consideration arises. The processing industry either has a limited capacity to process fish or sales and are limited by the market place. When these limits are reached the ex-vessel price per unit biomass will drop dramatically: the excess catch may be sold to bulk processors of sushimi (minced fish) or fish meal, or may be dumped. Thus in any quarter, the total biomass caught of species $j$ may be $B_j$ , say, while the actual biomass that can be marketed at a premium price is $B^0_j$ .

If $p_{1j}$ is the premium ex-vessel price per unit biomass for species $j$ then it follows from [11] (suppressing t and $\ell$) that the corresponding gross revenue $R_{rk}$ as a function of days $d_{rk\nu}$ spent per quarter on activity $\nu$ is

$$R_{rk\nu}(d_{rk\nu}) = \sum_{j=1}^{m_s} p_{1j}B_{jrk\nu} \qquad\qquad [21]$$

where

$$B_{jrk\nu} = \sum_{i=1}^{n_j} d_{rk\nu}\,\rho_{rk\nu}\,q_{ijk\nu}w_{ij}N_{ij}(1-e^{-(M_{ij}+F_{ij})/4})/(M_{ij}+F_{ij}) \qquad\qquad [22]$$

and $F_{ij}$ also dependent on $d_{rk}$ , is defined in [9].

The total revenue for boat $r$ in technology class $k$ across all activities is thus

$$R_{rk}(\underline{d}_{rk}) = \sum_{\nu=1}^{m_s+1} \sum_{j=1}^{m_s} p_{1j}B_{jrk\nu} \qquad\qquad [23]$$

provided the premium prices hold. Here $\underline{d}_{rk}$ is the vector of elements $d_{rk\nu}$ and must satisfy

$$\sum_{\nu=0}^{m_s+1} d_{rk\nu} = \bar{d}_{rk} \qquad\qquad [24]$$

where $\bar{d}_{rk}$ is the maximum number of days per quarter that boat $rk$ can expect to operate. This is usually less than the total number of days in a quarter since, due to inclement weather, one can expect to lose a certain number of fishing days each quarter, depending on the season. Note that we distinguish between the number of days we expect to be in port due to inclement weather and the decision variable $d_{rk0}$ (the number of days we choose to be in port).

Expression [23] only holds if the total biomass yield for species $j$ for the quarter is less than $B^0_j$ introduced above.

It follows from [22] that the total biomass of species $j$ caught firstly by boat $rk$ and secondly by the fleet as a whole respectively are,

$$B_{jrk} = \sum_{\nu=1}^{m_s+1} B_{jrk\nu} \qquad\qquad [25]$$

and

$$B_j = \sum_{k=1}^{m_f} \sum_{r=1}^{b_k} \sum_{\nu=1}^{m_s+1} B_{jrk\nu} \qquad\qquad [26]$$

If $B_j > B_j^o$ for some $j$, we will assume that the proportion

$$\gamma_j = B_j^o/B_j \qquad\qquad [27]$$

is disposed of at the premium price and a proportion $(1-\gamma_j)$ of the biomass is either sold at a lower price $p_{2j}$ or discarded, depending on the availability of an alternative market. In this case [23] becomes

$$R_{rk}(\underline{d}_{rk}) = \sum_{\nu=1}^{m_s+1} \sum_{j=1}^{m_s} (\gamma_j p_{1j} + (1-\gamma_j)p_{2j}) \, B_{jrk\nu} \qquad\qquad [28]$$

where

$$\gamma_j = 1 \quad \text{if} \quad B_j < B_j^o$$
$$\qquad\qquad\qquad\qquad\qquad [29]$$
$$\gamma_j = B_j^o/B_j \quad \text{if} \quad B_j > B_j^o$$

This analysis assumes that each boat operator sells the same fraction of its catch at the premium price. On average this is probably not true, but a refinement of this assumption is too complicated to deal with at this point.

Consider the optimal-activity-mix problem where boat $rk$ must find a $\underline{d}_{rk}$, satisfying [24] that maximizes

$$J_{rk}(\underline{d}_{rk}) = R_{rk}(\underline{d}_{rk}) - C_{rk}(\underline{d}_{rk}) , \qquad\qquad [30]$$

where $R_{rk}$ is defined in [28] and $C_{rk}$ is the combined cost across all activities (cf. the linear case in [20]), i.e.,

$$C_{rk}(\underline{d}_{rk}) = \sum_{\nu=1}^{m_s+1} C_{rk\nu}(\underline{d}_{rk}) \qquad\qquad [31]$$

This problem is insoluble unless $F_{ij}$ $j=1,\ldots,m_s$, $i=1,\ldots,n_j$ can be determined since, from [22], they are needed to determine the biomass quantities $B_{jrk\nu}$ and hence by [26] $B_j$ as well. From [14], however, the values of $d_{rk\nu}$ must be known for all $k=1,\ldots,m_f$, $r=1,\ldots,b_k$ and $\nu=1,\ldots,m_s$ before the $F_{ij}$'s can be determined.

In principle, the sole ownership problem of maximizing

$$J_s = \sum_{k=1}^{m_f} \sum_{r=1}^{b_k} J_{rk}(\underline{d}_{rk}) , \qquad\qquad [32]$$

subject to [24] holding for all r and k , can be solved using nonlinear programming techniques.

The n-player problem, where each of the boats belongs to one of n fleets, $F_i$ i=1,...,n, can be approached in one of several ways.

Define

$$J_i = \sum_{(r,k) \in F_i} J_{rk} \qquad i=1,\ldots,n \ .$$ [33]

then one of the following situations may exist.

(i)     Player i can maximize $J_i$ subject to the assumptions that the other players will most likely play strategies that follow known historical patterns.

(ii)    Player i=1,...,n play their nash equilibrium solution, i.e., their best strategies under the worst situation in terms of competition between players for the resource (c.f. Goh 1980).

(iii)   Players i=1,...,n agree to cooperate and a particular pareto-optimal solution is chosen, i.e., a solution which maximizes

$$J_\gamma = \sum_{i=1}^{n} \gamma_i J_i$$

where $\sum_{i=1}^{n} \gamma_i = 1$ , $J_i$ is given by [33], and constraints [24] are satisfied.

## Regulation of the fishery

Fisheries are regulated to preserve stocks from being overexploited with a view towards sustaining biomass yield over the long term and towards preventing the economic upheaval that fishing communities may be subjected to after a fishery has collapsed. These objectives can often be met by applying seasonal catch quotas either to individual species or to the multispecies fishery as a whole. In some multispecies fisheries, the productivity of individual species fluctuates strongly over a period of several years while the productivity of the fishery as a whole is more stable (Ursin 1982). In this case it may be more sensible to base regulations on total stock or catch levels. For purposes of illustration, the fecund stock level $S_j$ , defined in [5], will be used as a stock abundance index for the j-th species, although total stock or some other weighted sum of the number of individuals in each age class could be used. Thus, at time t , the total stock level for the fishery as a whole is (recalling [5])

$$S(t) = \sum_{j=1}^{m_S} S_j(t)$$ [35]

Suppose, for the moment, that escapement bounds $\overline{S}_j$ j=1,...,$m_S$ and $\overline{S}$ are known and it is desirable to regulate a fishery so that the total escapement at the beginning of each year, i.e., t=1,2,3, satisfies

$$S_j(t) > \overline{S}_j \qquad j=1,\ldots,m_S$$ [36]

and that

$$S(t) > \overline{S} \qquad\qquad [37]$$

If total stock regulation rather than individual stock regulation is emphasized then it will usually follow that

$$S(t) > \sum_{j=1}^{m_s} S_j \qquad\qquad [38]$$

The stock bounds $\overline{S}_j$ j=1,...,$m_s$ and $\overline{S}$ may be determined from long run simulations of the fishery at various exploitation levels where some subjective criterion is needed to evaluate the stock as overexploited (see Swartzman et al. 1983 for further discussion on this point.) The overall stock bound $\overline{S}$ could be set to encourage maximization over the long term of surplus production. The individual bounds $\overline{S}_j$ could be set at a level that prevents driving the j-th species to a point where it may collapse under a string of weak recruitment years. Note that the definition of a collapsed fishery is subjective.

In a competitive fishery, however, it is almost impossible to restrict catch directly by imposing seasonal quotas on each boat without destroying the competitive and capitalistic nature of that fishery. In extreme cases catch quotas for individual boats may be warranted. However, by limiting the number of days that a boat may participate in one or more activities during a season and by limiting entry of boats into the fishery (possibly by technology class) it is possible to indirectly limit catch rates. If the regulator knows the number of boats in the fishery and their fishing power ratings $\rho_{rk\nu}$ $\nu$=1,...,$m_s$, k=1,...,$m_f$, r=1,...,$b_k$ , then the activity levels $d_{rk\nu}$ can be limited by constraints $\overline{d}_\nu$ say, i.e.,

$$d_{rk\nu} < \overline{d}_\nu \qquad \nu =1,...,m_s , \qquad\qquad [39]$$

to ensure that $B_j$ in [26], j=1,...,$m_s$ , for the given quarter, does not exceed a desired catch level $\overline{B}_j$ , say. From [22] and [26] the values $\overline{d}_\nu$ must satisfy the equation

$$\sum_{\nu=1}^{m_s} \overline{d}_\nu \sum_{k=1}^{m_f} \sum_{r=1}^{b_k} \sum_{i=1}^{n_j} \rho_{rk\nu} q_{ijk\nu} w_{ij} N_{ij}(1-e^{-(M_{ij}+\overline{F}_{ij})/4})/(M_{ij}+\overline{F}_{ij}) = \overline{B}_j$$
$$j=1,...,m_s \qquad\qquad [40]$$

where, from [9],

$$\overline{F}_{ij} = \sum_{k=1}^{m_f} \sum_{r=1}^{b_k} \sum_{\nu=1}^{n_j} \overline{d}_\nu \, \rho_{rk\nu} q_{ijk\nu} \qquad\qquad [41]$$

In [40] we have $m_s$ nonlinear equations in $m_s$ unknowns $\overline{d}_\nu$ ,$\nu$=1,...,$m_s$ . If $\overline{F}_{ij}$ is sufficiently small the solution is unique, in general. In fact if $(M_{ij}+\overline{F}_{ij})/4$ is small compared with unity then an approximate solution to [40] can be obtained by replacing the expression between the left and right most parenthesis on the left hand side of [40] with (1/4) to obtain a linear system of equations in $\overline{d}_\nu$ .

If the fishery is regulated through constraints [39] then the fleet dynamics problems, defined in the previous section, of maximizing [30], [32] or [33] subject to [24] can be extended to a nonlinear programming problem that includes constraints [39].

Current regulations in our exemplar Oregon-Washington groundfish fishery restrict both trip frequency for a particular activity and catch per trip, but not length per trip or entry. Trip frequency restrictions only indirectly limit each $d_{rk\nu}$ while the lack of limited entry allows boats to enter the fishery unrestricted. Statistics indicate that the current policy fails to control catch rate.

Another important statistic for the regulator to monitor is the amount of biomass discarded. As mentioned in the previous section, biomass from the j-th species is discarded at a rate (cf. equations [26]-[29])

$$B_j^\delta = (1-\gamma_j) \sum_{k=1}^{m_f} \sum_{r=1}^{b_k} \sum_{\nu=1}^{m_s} B_{jrk\nu} \qquad j=1,...,m_s \qquad [42]$$

if $p_{2j}=0$ . Additional discarding may occur when less valuable fish are dumped from the hold to carry more valuable fish that are encountered on the same fishing trip.

If quotas are imposed to regulate catch then biomass over and above [42] will be discarded once the quota for the j-th species, say, is exceeded and the j-th species is caught as an incidental by-catch during some activity $A_\nu$, $\nu \neq j$ . This type of discard is difficult to monitor although it can be estimated if by-catch catchability coefficients are reasonably well known. One of the problems in the past is that discards have not been included in population analysis, such as VPA, that estimate population parameters. Thus population parameters estimated using landings as the sole source of catch data are invariably biased.

## Conclusion

Most fish species are exploited as part of a multispecies complex and in western countries multifleet situations are prevalent. The data needed to develop multispecies multifleet models must reflect technological interactions between boats and, where appropriate, biological interactions between species. If this type of data is unavailable, now, models still need to be conceptualized so that the appropriate data can be gathered to service these models in the future. The analysis presented here assumes that an extensive period during the year exists (in this case, one quarter) over which it is adequate to average out the behavior of individual boats and the stock response to fishing pressure. One alternative approach would be to model the day by day decision making process of a boat skipper or fleet owner. This, however, requires a much more extensive framework to characterize boats and possibly needs to consider the spatial distribution of the stock as well as a more detailed description of the economic structures of the fishery. Thus the framework presented here is seen as a first step in incorporating the fleet dynamics into a population dynamics model. Market forces are regarded as given inputs into the problem. A next step may be to incorporate market response and market dynamics into the model by coupling it with catch rate and some notions of supply, demand and capitalization of the processing industry. The question of entry and exit of boats into the fishery also needs to be dealt with as does the application of the model to questions relating to long term exploitation, i.e., over a five to twenty year time horizon.

# REFERENCES

Beverton, B. J. and S. J. Holt, 1957. On the Dynamics of Exploited Fish Populations, Fisheries Investigation Series II, Vol. 19, Ministry of Agriculture, Fisheries and Food, London.

Clark, C. W. 1976. Mathematicial Bioeconomics, Wiley-Interscience, New York.

Clark, C. W. and G. R. Munro. 1980. Fisheries and the processing sector: Some implications for management policy, Beil. J. Econ., 11: 603-616.

Deriso, R. B., 1980. Harvesting strategies and parameter estimation for an age structured model. Can. J. Fish. Aquat. Sci. 37: 268-282.

Getz, W. M., 1979. Optimal harvesting of structured populations, Math. Biosci. 44: 269-291.

Getz, W. M., 1980a. Harvesting models and stock recruitment curves in fisheries management, in Mathematical Modelling in Biology and Ecology (W. M. Getz, Ed.), Springer, Heidelberg.

Getz, W. M., 1980b. The ultimate sustainable-yield problem in nonlinear age-structured populations, Math. Biosci. 48: 279-292.

Goh, B. S. 1980. Management and Analysis of Biological Populations, Elsevier, Amsterdam.

Getz, W. M., 1984. Production models for nonlinear stochastic age-structured fisheries, Math. Biosci. 69: 11-30.

Laevastu, J., F. Favorite, and H. A. Larkins, 1982. Resource assessment and evaluation of the dynamics of the fisheries resources in the northeastern Pacific with numerical ecosystem models. in Mercer (ed.) 1982 (op. cit): 70-81.

Levin, S. A. and C. P. Goodyear, 1986. Analysis of an age-structured fishery model, J. Math. Biol. 9: 245-274.

Murawski, S. A., 1984. Mixed-species yield-per-recruitment analyses accounting for technological interactions. Can. J. Fish. Aquat. Sci. 41: 897-916.

May, R. M., J. R. Beddington, C. W. Clark, S. J. Holt, R. M. Laws, 1979. Management of multispecies fisheries. Science 205: 267-277.

McKelvey, R., 1982a. Economic regulation of targeting behavior in a geographically extensive multispecies fishery. Interdisciplinary Series Report No. 19A, Dept. of Mathematics, University of Montana, Missoula.

McKelvey, R., 1982b. The fishery in a fluctuating environment: Coexistence of specialist and generalist fishing vessels in a multipurpose fleet. Interdisciplinary Series Report No. 20, Dept. of Mathematics, University of Montana, Missoula.

Mercer, M. C. (editor), 1982. Multispecies approaches to fisheries management advice. Can. Spec. Publ. Fish. Aquat. Sci. 59.

Pope, J. G., 1972. An investigation of the accuracy of virtual population analysis usin cohort analysis, Int. Comm. Northwest Atl. Fish. Res. Bull. 9: 65-74.

Reed, W. J., 1980. Optimum age-specific harvesting in a nonlinear population model, Biometrics 36: 579-593.

Swartzman, G. L., W. M. Getz, R. C. Francis, R. Haar, and K. Rose., 1983. A management analysis of the Pacific whiting fishery using an age structure stochastic recruitment model, Canad. J. Fish. Aquat. Sci. 40: 524-529.

Ursin, E., 1982. Multispecies fish stock assessment in ICES. In Mercer (ed.) 1982, (op. cit.): 39-67.

# RISK ADVERSE HARVESTING STRATEGIES

Richard Deriso
International Pacific Halibut Commission
P.O. Box 95009
Seattle, Washington  98145-2009

## Abstract

Results are presented on harvest strategies that optimize a risk adverse management objective.  Risk in this paper refers to a reluctance on the part of managers to
freely gamble with current quotas in an attempt to increase the odds for a higher
future catch.  Harvest policies that maximize a logarithmic utility function, one type
of risk adverse objective, are shown to differ substantially from the fixed escapement
policies that maximize average catch.  In particular, a constant harvest rate policy
maximizes log-catch in one model.  A number of advantages to constant harvest rate
strategies are discussed.

## Introduction

Fixed escapement policies are optimal harvest strategies for a large class of
population models where the objective is to maximize some risk neutral function of
catch (Reed, 1979).  A fixed escapement policy is characterized by an optimal target
escapement $S_{opt}$, independent of the current recruitment level $R$, and the optimal harvest is given by max($R - S_{opt}$, 0).  The risk neutral feature of the management policy
is perhaps not widely appreciated as being critical to the conclusion of optimality of
fixed escapement.  By a risk neutral policy, I mean one where the one-period value or
utility of catch is a linear function of catch and it is characterized by having a
constant first derivative (see for example Hilborn and Walters, 1977).  Risk neutral
also implies that the management policy is indifferent to a given catch now versus
gambling on a future catch that is equal in expected value (even if there is a significant probability of no future catch).  The gamble is shown clearly in Figure 1 where
the probability of shutting down the fishery (pr. $R < S_{opt}$) is plotted for the optimal
escapement policy that maximizes average catch in the Ricker spawner-recruit model,

$$R = \exp(a - bS + W) \tag{1}$$

where $R$ is recruitment, $S$ is escapement, $a$ and $b$ are parameters, and $W$ is a normally
distributed $N(0, \sigma^2)$ random variable.  For example, in Figure 1 we see that a fishery
will not be allowed to operate in 10% of the years for a stock of moderate productivity
($a = 1.2$) and moderate stochasticity ($\sigma = 0.5$).

Risk adverse utility functions are characterized by a decreasing marginal utility
and a reluctance on the part of managers to freely gamble with current quotas in an
attempt to increase the odds for a higher future catch.  This would mean that a unit
of biomass in a small catch is worth more than the same unit in a larger catch.  This

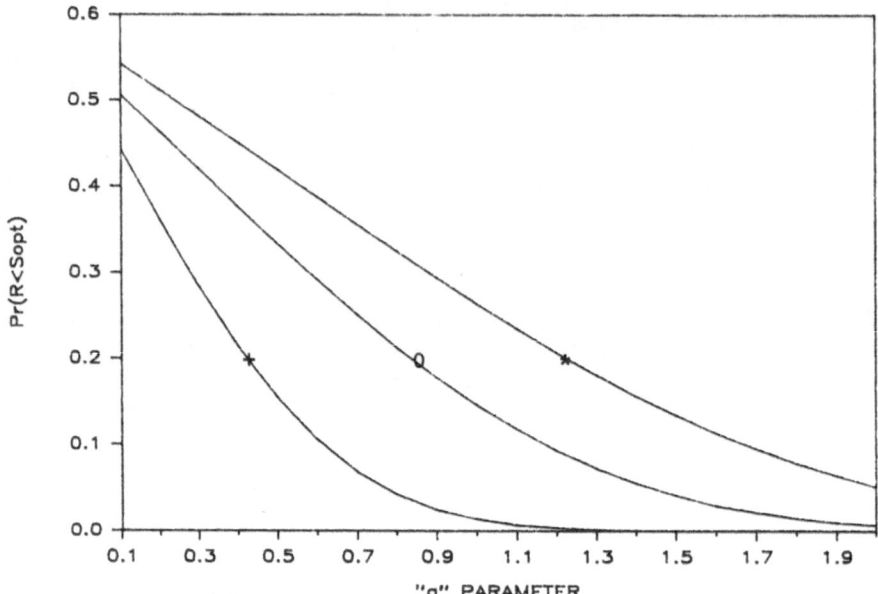

Figure 1. Probability that recruitment is less than the optimal escapement level for
the Ricker model (1) under three levels of natural variability; $\sigma = 0.25$ for curve (+),
$\sigma = 0.5$ for curve (0), $\sigma = 0.75$ for curve (*).

makes sense from the viewpoint of a fishery biologist, a fishery manager, or a commercial fisherman. Here the stability of catch is important along with the amount of
catch. IPHC Director, D.A. McCaughran, speaking about the current management objectives for the Pacific halibut fishery, states in Deriso (1985), "We still seek to maximize sustainable yield, but the emphasis now is on sustainable and not necessarily
maximum. We try to avoid causing a boom and bust type fishery, as has occurred in the
past." Dr. McCaughran is typical of most fisheries decision makers, I believe, in
being normally risk adverse.

Stock portfolio managers have long realized the advantages of a <u>logarithmic utility function</u> (Thorp, 1975) for risk adverse policies and we are beginning to see its
usage in fisheries analysis (Mendelssohn, 1982; Ruppert et al., 1985). Thus, an alternative to maximizing "arithmetic" average catch is to maximize the expected value
of the geometric mean of catch or equivalently,

$$\text{maximize } E \left( \sum_{t=1}^{N} \ln(C_t) \right) \tag{2}$$

where $C_t$ is the catch in year $\underline{t}$ out of a total of $\underline{N}$ years. This objective function is
similar to ones considered in both the previously cited fisheries papers. Both of
those papers show that solutions for such an objective allow some catch to occur at
all recruitment levels, and produce a smoother sequence of catches.

## Analysis of logarithmic objective functions

Constant harvest rate policies are one way to produce a reasonably smooth sequence of catches. In these policies, catch is proportional to annual abundance. Basically, this type of policy will split the natural stochasticity of the system between the fishery and the fish stock. In contrast, a constant escapement policy marks one extreme which forces the fishery to absorb all the randomness in the system in the form of fluctuating catches. The other extremes are constant catch policies that can produce cautious and stable catches although the fish stock absorbs all the natural stochasticity.

A constant harvest rate policy maximizes objective (2) for large N and for a population modeled with a recruitment curve of the form

$$R = e^{a+W}S^b \tag{3}$$

(Cushing and Harris 1973) where $\underline{a}$ and $\underline{b}$ are parameters and W is a normally distributed random variable.

This result is easy to show using stochastic dynamic programming since the dynamic equation (3) under a logarithm transform is a semilinear equation and since the objective (2) is a semilinear equation (Bertsekas 1976, page 67). In other words, the dynamic system is of the form

$$X_{t+1} = AX_t + f(U_t) + W_t$$

where $X_t$ is the state variable ($\ln(R_t)$ in our case) and $f(U_t)$ is a function of the control applied at time t ($\ln(1-\mu_t)$ where the harvest rate $\mu_t$ is chosen as the control function in our case). The objective function (2) is of the semilinear form

$$E \left( \sum_{t=1}^{N} X_t + g(U_t) \right)$$

with $g(U_t) = \ln(\mu_t)$ in our case. For semilinear systems, if there is an optimal policy then it consists of constant functions $U^*_t$, i.e. $U^*_t(X_t)$ = constant for all $X_t$, and for large N these policies usually approach a stationary policy (which certainly occurs in our case).

Constant harvest rate policies are not always optimal for maximizing (2), the logarithmic catch sum. A good example is given in Mendelssohn (1982) which shows the optimal logarithmic policy for a specific Ricker recruitment curve allows an approximate fixed escapement when recruitment is high. The declining right-hand limb, characteristic of dome-shaped recruitment curves such as the Ricker model (1), is the reason an optimal logarithmic policy would limit the amount of spawning to be less than some upper bound. The optimality equation in dynamic programming can be used to illustrate properties of a logarithmic utility function. Optimal stationary policies are characterized by

$$J*(R) = \max_{S} E(\ln(R-s) + J* (F(S,W)))  \qquad (4)$$

where J* is the optimal cost-to-go function and the control function U(R) is chosen
as S(R), the escapement level for future spawning in the recruitment function F(S,W).
The necessary condition for an interior optimum is

$$0 = \frac{-1}{R-S} + E\left(\frac{dJ*}{dF} \frac{dF}{dS}\right)$$

but $\frac{EdJ*}{dF}$ (F) can be found by differentiating (4), evaluated at F(S,W), to get

$$E \frac{dJ*(F)}{dF} = E \quad (F(S,W) - S(F(S,W)))^{-1}$$

In order to examine the qualitative features of the optimum, replace all random quantities by their expected values to get the corresponding deterministic optimum

$$\frac{F(S) - S(F(S))}{R - S(R)} = F'(S)$$

The left-side is just expected catch next year divided by actual catch this year,
which is a non-negative number.  If F' < 0 for large S values (i.e. the recruitment
function is dome-shaped) then S(R) clearly approaches the escapement level that maximizes F(S) as R increases.  The important conclusion here is that the shape of the
recruitment curve qualitatively determines the type of policy optimal for objective
(2).  Except in instances where over-spawning causes recruitment failure, the constant
harvest rate policies that I have examined in numerical experiments produce high and
relatively stable catches.


Reliability of parameter estimates

Parameter identification is another reason one may prefer fixed harvest rate policies over constant escapement.  Escapement is clearly a quantity dependent upon the
size of carrying capacity for each specific species.  Optimal harvest rates are independent of the carrying capacity type parameter in a large family of both age-structured
and non-age-structured population models.  Specifically, the harvest rate that maximizes equilibrium catch is independent of parameter k in all recruitment models of the
form,

$$R = Sf(S/k),  \qquad (5)$$

where the parameter k only occurs in the model as a divisor (or multiplier) of spawning stock.  To prove this result, differentiate (5) with respect to S and set the derivative to one and get

$$f(\chi) + \chi f'(\chi) = 1$$

where $\chi = S/k$, the level of relative optimum stock size, is determined implicitly in
the above equation.  Optimal harvest rate $\mu*$ is then given by

$$\mu^* = 1 - \frac{1}{f(\chi)},$$

which is independent of the parameter k.

All the commonly used recruitment models are in the forms (5), including the Beverton and Holt (1957) curve and the Ricker (1954) curve. For these two models, optimal harvest rate is given by

$$\mu^* = 1 - \sqrt{\alpha} \qquad \text{for model } R = \frac{S}{\alpha + \beta S}$$

$$\mu^* \cong .5a - .07a^2 \qquad \text{for model } R = S \exp(a - bS)$$

where that approximation for the Ricker model is reasonable for a values in the range $0 < a < 2$. In the logistic production model optimal fishing mortality is given by

$$F^* = r/2$$

where r is the intrinsic rate of growth. Note the similarity between optimal harvest rates for the Ricker and logistic model: both rates are around one-half the density-dependent production rate (given as a and r in those models, respectively).

Any situation in which we are relatively certain about the intrinsic productivity rate of a fish stock but uncertain about the size of carrying capacity would mean we had good confidence in our estimates of optimal harvest rate, but little confidence in our estimates of the optimal escapement. The major circumstance in which we might be confident about the productivity but uncertain about the size is when a stock is overexploited, a frequent occurrence. When a stock is overexploited, we have a good idea of its intrinsic productivity because the stock is low enough to reduce density dependence, yet we do not know how large the stock would have to be to start showing reduced rates of return.

The Ricker recruitment model provides a good framework to examine the relationship between observed stock abundance and variance of model parameter estimates. The logarithmically transformed model (1) becomes

$$\ln(R/S) = a - bS + w,$$

a standard linear regression format with y variable, $\ln(R/S)$, and x variable, S. Maximum likelihood estimates for a and b, obtained in the usual manner, have variances

$$\text{Var}(\hat{b}) = \frac{\sigma^2}{\Sigma(S_i - \overline{S})^2}$$

and $\text{Var}(\hat{a}) = \sigma^2 \left( \frac{1}{N} + \frac{\overline{S}^2}{\Sigma(S_i - \overline{S})^2} \right)$

where the dispersion of observed stock abundance about mean stock level $\overline{S}$ is given by $\Sigma(S_i - \overline{S})^2$. Variance of the productivity rate, $\text{Var}(\hat{a})$, decreases with the square of mean stock, $\overline{S}$, for a fixed dispersion of observed stock abundance and for small $\overline{S}$ we have

$$\text{Var}(\hat{a}) \cong \frac{\sigma^2}{N} \, .$$

The variance associated with the density-dependent coefficient, $\hat{b}$, does not have this effect and, in fact, it is invariant to change in $\bar{S}$, as long as the dispersion of stock abundance stays constant.

In most fisheries, stock abundance is not known precisely but rather it is an estimate subject to possibly large measurement error. Such errors cause fisheries management to set catch quotas that can differ dramatically from optimal catch; the difference is dependent upon the type of harvest policy employed. Let us look at a typical scenario in fisheries management: Standing stock prior to harvest, say recruitment R for a single-age fishery, is estimated as $\hat{R}$. The catch quota is then set according to a specified harvest policy. We are interested in determining how close the catch quota is to the optimum catch that would be taken if we knew the true recruitment level. An appropriate measure of that precision is the variance of a given catch, Var(Catch), where variability is measured about the measurement error $\hat{R}$ and, possibly, in the harvest policy parameters as well.

When harvest policy parameters are assumed known without error then the Var(Catch) arises only from measurement error in the standing stock estimate, $\hat{R}$. For a fixed harvest rate policy with constant exploitation fraction $\mu$,

$$\text{Var(Catch)} = \text{Var}(\mu\hat{R}) = \mu^2 \, \text{Var}(\hat{R})$$

is the variance in the catch quota. For a fixed escapement policy with escapement level S*,

$$\text{Var(Catch)} = \text{Var}(\hat{R} - S*) = \text{Var}(\hat{R})$$

is the variance in catch quota.

The harvest rate policy is more precise than the escapement policy when policy parameters have been decided upon but standing stock is an estimate. The equations above show that the variance of catch from a constant harvest rate policy is $\mu^2$ times the Var(Catch) for an escapement policy; for example, with a 20% harvest rate, the Var(Catch) for a constant harvest rate policy is 4% the Var(Catch) for a fixed escapement policy. The effects of this high variability in catch are superimposed on top of what was already highly fluctuating catches by the nature of fixed escapement policies (as discussed in the introduction). Such high variability has at least two effects. First, true escapement will vary from year to year despite attempts to manage on a fixed escapement basis. Secondly, the variability in true escapement implies an increase in the probability of overexploiting the stock, as measurement error increases.

Analysis becomes more complicated when optimal harvest ($\mu$) and optimal escapement (S*), as well as standing stock (R), are all estimates subject to error. For a fixed harvest rate policy,

$$\text{Var(Catch)} = \text{Var} \; (\hat{\mu}\hat{R}) \tag{6}$$

$$= \hat{\mu}^2 \; \text{Var}(\hat{R}) + \hat{R}^2 \; \text{Var}(\hat{\mu}) + \text{Var}(\hat{R}) \;\; \text{Var}(\hat{\mu})$$

is the variance in catch quota arising from uncertainty in the estimates, R and μ. For a fixed escapement policy,

$$\text{Var(Catch)} = \text{Var}(\hat{R} - \hat{S}*) \tag{7}$$

$$= \text{Var}(\hat{R}) + \text{Var}(\hat{S}*)$$

is the variance in catch quota arising from uncertainty in the estimates $\hat{R}$ and $\hat{S}*$. Comparing (6) to (7) we see that a sufficient condition for a fixed harvest rate policy to be the more precise of the two, is

$$\hat{\mu}^2 + \text{Var}(\hat{\mu}) < 1$$

and $\hat{R}^2 \; \text{Var}(\hat{\mu}) < \text{Var}(\hat{S}*)$,

which is usually the case. Exceptions include when $\hat{\mu}$ is far less precise than $\hat{S}*$ and when $\hat{R}$ is very large compared to $\hat{S}*$.

## Discussion

Fisheries management objectives are usually stated in terms of "maximum benefits" that provide little guidance to managers. In many fisheries, there has been a distinct movement away from the maximum biological production to maximizing a mix of average benefits and reduced year to year variation. The distinction between linear and logarithmic utilities discussed in this paper captures this difference. A linear utility will be most appropriate when the fishery constitutes a small portion of the fishermen's annual income and where price and costs are relatively unaffected by volume of harvest. The logarithmic utility will be appropriate when the fishery constitutes the major portion of the fishermen's income or price and costs vary with volume. If stability of volume is critical for marketing reasons, then a logarithmic utility will be more appropriate than a linear utility.

Actual fisheries management practice illustrate circumstances under which different utilities are appropriate. In the east coast cod fisheries of Canada, it was recognized at the declaration of extended jurisdiction that the stocks were depressed and severely overfished. Management response was not to cease fishing until the stock was rebuilt, but to adopt the $F_{0.1}$ policy (a fixed harvest rate) to allow a fishery and rebuild the stocks simultaneously. Similarly, in the late 1970's, the International Pacific Halibut Commission determined that the halibut stock was depressed and needed to be rebuilt. The IPHC did not adopt a fixed escapement policy, but instead allowed the stock to rebuild more gradually, allowing harvests to increase as the stock increased. In both of these fisheries, the fishermen were largely dependent upon the stock. There are numerous cases where fishing has been halted to allow a stock to rebuild; these include the Rivers Inlet sockeye salmon in British Columbia where many other salmon fisheries were available to the fishermen, the Pacific herring fishery in B.C. in the late 1960's, and the Alaska Fur Seal Fishery at the turn of the century.

This paper covers just a few of the issues concerning risk in fisheries management policy. Complications in the dynamics of fish (e.g. multi-species, multi-users), as well as complications in the measurement of fish abundance can radically alter the form of the optional harvest policy (see for example, Walters 1981; Clark and Kirkwood 1984). The probability of quasi-extinction of a population is another type of risk not covered explicitly by utility functions. The unbounded negative range for a logarithm near zero does cause a low harvest of small stocks in logarithm catch policies, but generally they do not offer the level of protection given by a fixed escapement policy that completely closes a fishery when recruitment is below some optimum amount. A policy that is both utility risk adverse and adverse to stock quasi-extinction may be desirable. One such policy that can be applied in fisheries management is to set an absolute minimum stock size that triggers a closing of the fishery and a fixed harvest rate in a wide middle range of stock sizes.

## Acknowledgements

I thank Ray Hilborn for ideas and discussions on current management practices and on fixed harvest rate policies. I am grateful to IPHC for supporting this research.

Bibliography

Beverton, R. J. H. and S. J. Holt. 1957. On the dynamics of exploited fish popula-
    tions. U.K. Min. Agric. Fish., Fish. Invest. (Ser. 2) 19:533 pp.

Bertsekas, D. P. 1976. Dynamic Programming and Stochastic Control. Academic Press,
    New York. 397 p.

Clark, C. W. and G. P. Kirkwood. 1984. On uncertain resource stocks: optimal harvest
    policies and the value of stock surveys. Tech. Rpt. No. 84-4, The Institute of
    Applied Mathematics and Statistics. University of British Columbia, Vancouver,
    B.C.

Cushing, D. H. and J. G. K. Harris. 1973. Stock and recruitment and the problem of
    density-dependence. Rapp. P.-V. Reun. Cons. Explor. Perm. Int. Mer. 164:142-155.

Deriso, R. B. 1985. Stock assessment and new evidence of density-dependence. (In)
    Fisheries Dynamics: harvest, management, and sampling. Washington Sea Grant, WSG
    85-1, Seattle, WA, pp. 49-59.

Hilborn, R. and C. J. Walters. 1977. Differing goals of salmon management on the
    Skeena River. J. Fish. Res. Board Can. 34:64-72.

Mendelssohn, R. 1982. Discount factors and risk aversion in managing random fish
    populations. Can. J. Fish. Aquat. Sci. 39:1252-1257.

Reed, W. J. 1979. Optimal escapement levels in stochastic and deterministic harvest-
    ing models. J. Environ. Econ. Manag. 6:350-363.

Ricker, W. E. 1954. Stock and recruitment, J. Fish. Res. Board Can. 11:559-623.

Ruppert, D., R. L. Reish, R. B. Deriso, and R. J. Carroll. 1985. A stochastic popu-
    lation model for managing Atlantic menhaden (Brevoortia tyrannus) and assessing
    managerial risks, Can. J. Fish. Aquat. Sci: to be published.

Thorp, E. O. 1975. Portfolio choice and the Kelly criterion. (In) Stochastic Optimi-
    zation Models in Finance. W. T. Ziemba and R. G. Vickson Eds. Academic Press,
    New York.

Walters, C. J. 1981. Optimum escapements in the face of alternative recruitment hy-
    potheses. Can. J. Fish. Aquat. Sci. 38:678-689.

# A COMPARISON OF HARVEST POLICIES FOR MIXED STOCK FISHERIES

Ray Hilborn
Institute of Animal Resource Ecology
University of British Columbia
Vancouver, B.C.V6T 1W5, Canada

## Abstract

Existing theory on optimal harvesting policies for exploited populations deals almost exclusively with single stocks of known productivity managed for maximization of discounted average catch. This paper examines alternative harvest policies for mixed stock fisheries in which the manager cannot distinguish between abundances of separate stocks and can deal with total abundance only. Numerical methods are used to show that if the stocks have uncorrelated natural variation, the best harvest policy will be very much like a constant harvest rate rather than a constant escapement. When natural variation is correlated between stocks, even if productivities of stocks are different, a fixed escapement policy appears to maximize average catch. When the objective is not maximization of average catch, but rather the sum of logarithms of the catch, the best policy appears to be quite similar to constant harvest rate.

## 1.  Introduction

The literature on optimal harvest policies for exploited populations has focused almost exclusively on the simplest starting assumptions, a single stock of known productivity harvested to maximize average catch (1,2,3,4). A number of authors have developed various extensions of the basic model using discounted average catch and allowing for various forms of stochastic dynamics (5,6). The results of this work have shown that the optimal harvest policy is one that holds the population as closely as possible to a specific optimum size. When the population is below this level, no fishing takes place; when the population is above this level, it is harvested as quickly as possible to bring it down to optimum size. Such policies are often called fixed escapement policies, where escapement refers to the number of individuals allowed to escape from harvesting. The term originally applied to management of Pacific salmon, but now "fixed escapement harvest policy" is used to refer to any policy which attempts to hold the population at a constant size.

Recent work has explored the consequences of uncertainty in the parameters of the production model (7,8,9,10,11) and found that policies which use information about uncertainty of parameter estimates will outperform policies which do not use this information.

Other authors have explored the consequences of mixed stock fishing (12,13), but in these cases the authors assumed that the manager could determine the abundance of the individual component stocks. Recently Clark and Kirkwood (14) have examined the consequences of uncertainty in determination of stock abundance for a single stock.

In practice, few if any fisheries deal with a single stock of uniform productivity with known stock size and production parameters and with an objective of maximum discounted average catch. In other words none of this theory applies to most of the real world. In this paper I present some numerical explorations of the consequences of mixed stocks and management objectives other than maximum average catch. In particular, I explore management of stocks that consist of numerous substocks with separate dynamics and whose interannual variation may be correlated or uncorrelated. I explore cases where these stocks have the same intrinsic productivity and cases where productivities are different.

## 2. Forms of Harvest Policies

The two most frequently discussed harvest policies are constant harvest rate policies in which a fixed proportion of the population is harvested each year and a fixed escapement policy described earlier. Other types of harvest policies which have been discussed include probing policies where the harvest depends upon aspects of stock dynamics other than stock size. All policies that consider the catch only as a function of the current stock size (these include fixed harvest rate and fixed escapement) can be expressed graphically in a simple way as illustrated in figure 1. Shown in addition to fixed escapement and fixed harvest rate is a constant catch policy; one that has little history of implementation or theoretical analysis due to limitations when stock size becomes small.

Note that the three policies described above are straight lines in figure 1, thus they can be described by the following equation:

$$\text{Catch} \quad = \quad \text{intercept} \quad + \quad \text{slope x stock size} \qquad (2.1)$$

Constant harvest rate policies have a slope equal to the harvest rate and an intercept of zero. Constant catch policies have a slope of zero with an intercept equal to the catch, whereas fixed escapement policies have a slope of 1.0 and an intercept equal to minus the desired escapement.

Obviously more complex policies could be examined that would require higher order terms, but for this initial exploration we will confine ourselves to policies that can be described by a slope and an intercept. Using equation 2.1 we can postulate a specific biological model of the harvested stock and then search for the parameters of equation 2.1 that maximized the specific objective. This type of optimization is called fixed form optimization and is described in (8). If the postulated biological model is simple enough (3 or fewer state variables), stochastic dynamic programming could be used to find the optimal policy. However, the purpose of this paper is to explore and compare alternative harvest policies under various assumptions about fish biology and management objectives. Therefore I am less concerned with finding the true optimal policy than with examining the consequences of different policies.

In section 3 I will describe a number of rather simple mixed stock biological models and then in section 4 examine the performance of harvest strategies defined by equation 2.1 when applied to these different models.

## 3.   Biological Models

I assume that the harvested stock consists of a number of discrete stocks that happen to mix at the point of harvest and that the manager cannot discriminate between stocks. I assume that the manager has a perfect measure of the abundance of the total stock. The stock dynamics are assumed to be governed by a stock recruitment relationship of the Ricker type:

$$N_{t+1} = N_t \text{ x exp(a (1.0 - } N_t\text{/B) ) x exp(w)} \qquad (3.1)$$

where N is the population size, a is a measure of productivity, B is a scaling parameter (and the unfished equilibrium stock size) and w is a random error process that is normally distributed with a mean of zero and some variance.

The two major biological issues in mixed stock harvesting are

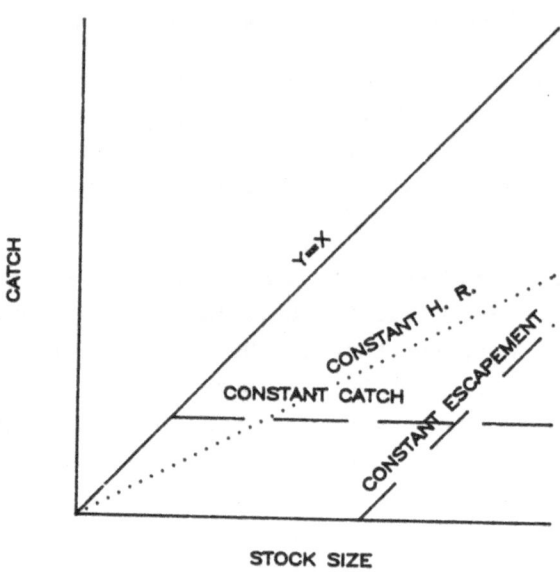

Figure 1. Catch vs stock size for fixed escapement, constant harvest rate and constant catch policies.

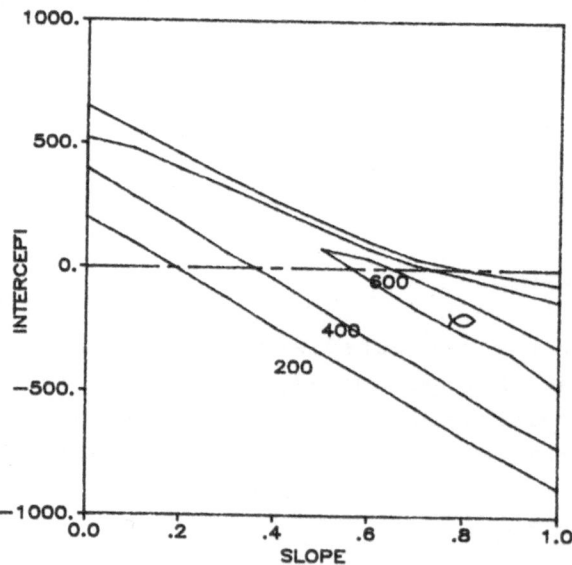

Figure 2. Average annual yield contours for different policies defined by a slope and an intercept for mixed stocks with the same productivities and uncorrelated random variation.

whether the stocks have the same productivity and whether the year-to-year natural variation (process error) is correlated. The optimum harvest rate depends primarily upon the a parameter (to a lesser degree on the variance of w), and if all stocks have the same productivity, then the implications for mixed stock harvesting may be quite different than for a wide range of productivities. Similarly, the correlation of year-to-year variation is important. For instance, if the stocks are perfectly correlated and have the same productivity, we are effectively looking at one stock. However, if the stocks are uncorrelated in their variation, determination of a good harvesting plan may be more difficult.

Since the case of same productivities and perfectly correlated variation is essentially the same as the well studied single stock case, I will ignore it here. Instead, I will examine three new cases: 1) same productivity and uncorrelated variation, 2) different productivity but correlated variation, and 3) different productivity and uncorrelated variation.

For all simulations I used 10 stocks, each with a B value of 100 and the standard deviation of w equal to 0.5. When productivities were the same, I used a=1.5; when productivities were different I used a=1.0 to 1.9 in steps of 0.1.

## 4. Results

For each biological model 500 different harvest policies were tested. Harvest policies were a slope (0.0 to 1.0 in units of 0.1) and an intercept (-1000 to 1000 in units of 40) as in equation 2.1. The biological model was simulated for 30 years using equation 3.1 for each stock. The total abundance of all stocks was combined to determine the size of the aggregate stock, and then the catch prescribed by the harvest policy was removed from each stock in proportion to its abundance. For each harvest policy the 30-year simulation was repeated 100 times and the average annual yield used. In fact, 5 to 10 Monte Carlo repetitions provided almost exactly the same answers.

Three measures of performance were calculated. The first measure is the traditional annual catch. The second measure was the average natural logarithm of the yearly catch. 1.0 was added to the catch each year so that we would never have to take the logarithm of 0.0. The third measure was the coefficient of variation of the annual catches. The logarithmic measure of performance is a surrogate for a declining marginal value of increasing catch and reflects what may be a more realistic objective for many of the world's fisheries. The declining marginal value of catch may be due to economic or social reasons. Deriso's paper in this volume discusses this concept in more detail.

Figure 2 shows the results from case (1), same productivity but uncorrelated errors. It is worthwhile devoting some time to understanding the structure of this type of graph since the next three figures use the same mode of presentation. The isopleths are the contours of equal annual yield; shown here are the contours for 200, 300, and 400. The fish symbol marks the harvest policy that had the highest average catch. The three reference type of policies, fixed escapement, fixed harvest rate, and constant catch can be traced on figure 2 as follows.

The right-hand vertical edge of figure 2 (slope=1.0) represents all possible fixed escapement strategies. The lowest point (intercept=-1000) represents a fixed escapement policy of 1000 units. Since we are dealing with 10 stocks, each with a B value of 100, we would expect the total stock size to exceed 1000 quite rarely and as we see, a fixed escapement policy of 1000 rarely harvests fish. Going up the right-hand edge of figure 2, we are moving to lower and lower escapement targets and we see the yield increase until it reaches a maximum in the 300-500 range and then decreases as escapements produce smaller and smaller yields.

The optimum escapement for a single stock with a=1.5 and B=100 is 39.5, therefore the optimum for 10 identical stocks would be 395. However, the lack of correlation in the errors and the presence of year-to-year variation modify the best escapement to some extent. When we get half-way up the right-hand edge of figure 2 we reach a fixed escapement policy with an escapement of 0. This provides little room for sustainable catch.

Fixed harvest rate policies can be traced along the dashed horizontal line in the middle of the graph. This dashed line represents all slopes that have an intercept of 0.0. The slope represents the harvest rate. Thus with a harvest rate of 0.0, there is no yield, but as we increase harvest rate up to about 0.5 we see yield rise, and then start to decline.

All possible constant catch policies can be traced by starting half-way up the left-hand edge (slope=0,intercept=0), and then continuing upward. A constant catch policy will perform reasonably well if the catch required is small enough, but once the annual desired catch gets very large, all the fish are taken in one year and the stock is gone. Since the results on this graph are the average annual catch during 30 years of simulation (and not equilibrium yield), a quota policy that drives the stock to extinction still produces some catch during the 30-year simulations.

The best policy in this figure is neither fixed escapement nor fixed harvest rate, but rather somewhere in between. Of more significance however is the long ridge of good harvest policies; obviously many policies, ranging from fixed escapement to fixed harvest rate, perform almost equally well. In terms of average yield, there is little to choose among them.

Figure 3 shows the average logarithm of catch for the same simulation as in figure 2. Again the fish symbol marks the best policy and the contours show policies of equal value. The best policy is very close to being a constant harvest rate. Certainly no fixed escapement policy performs particularly well for logarithmic objectives, because of the high year-to-year variability generated by fixed escapement policies. Again, there is a ridge of near equal policies, and these range from a fixed harvest rate policy to some that have steeper slopes and do not harvest at very small stock sizes. Logarithmic objectives place a premium on having as few years as possible without fishing, and this is achieved by having intercepts rather close to 0.0.

Figure 4 shows the average yield isopleths for the case of different productivities and perfectly correlated variation. In this case the best policy is a fixed escapement policy with a ridge leading up to a constant harvest rate of 0.5. Constant quota policies do not perform particularly well because when errors are correlated all stocks have good and bad years at the same time, and therefore the total stock will frequently be rather small. When errors are uncorrelated, the chance that all stocks will have a poor year is very low. The fixed harvest rate policy performs reasonably well, although the fixed escapement policy does produce a higher average yield. The logarithmic utility is not shown but looks very similar to figure 3. Again the optimum is more like a fixed harvest rate than a fixed escapement.

Figure 5 shows the results for the last case in which the stocks

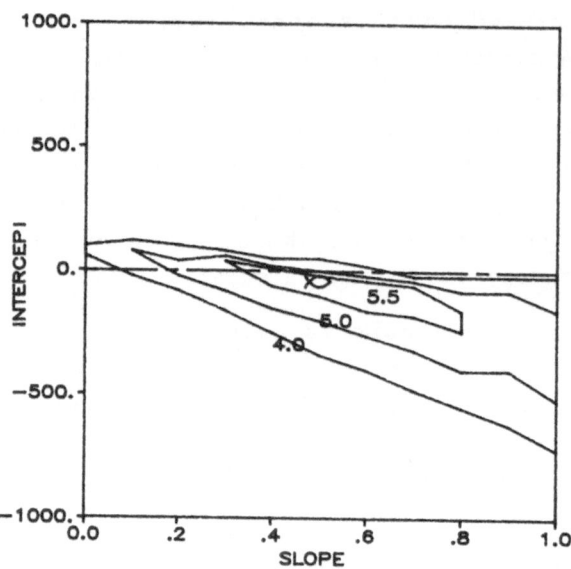

Figure 3. Average logarithm of annual yield contours for different policies defined by a slope and an intercept for mixed stocks with the same productivities and uncorrelated random variation.

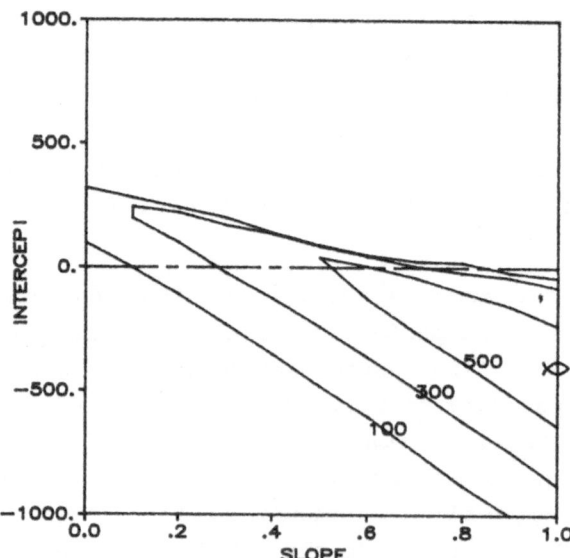

Figure 4. Average annual yield contours for different policies defined by a slope and an intercept for mixed stocks with different productivities and correlated random variation.

have different productivities and uncorrelated variability. The results are similar to figure 2, the best policy is between a fixed escapement and fixed harvest rate. However, because stock productivities vary, the yield of the aggregate is dominated by the yield from the few highly productive stocks. Therefore the optimum harvest rate is higher and the optimum escapement is lower than for a case of equal productivities. Hilborn (15) discusses this in more detail.

It is obvious that there is a trade-off between the average yield and the variability in yield. For single stock fisheries fixed escapement policies maximize average yield and also maximize the variance of the yield. The most desirable policy for any fishery will depend on how willing the manager is to trade reduced variability for lower average yield. We can see above that in mixed stock fisheries with uncorrelated variability the policy that maximizes average yield is not fixed escapement and has lower variability than a fixed escapement policy. Nevertheless there remains a trade-off between average yield and variability. We can illustrate this as in figure 6. This figure shows the range of average yields and coefficients of variation in yield that can be obtained by all possible combinations of slope and intercept of equation 2.1, operating on the mixed stocks with the same productivities and uncorrelated errors. The highest possible yield is over 600 and is obtained at many points on the ridge shown in figure 2. However, the left-hand end of this ridge has lower coefficients of variation than the right-hand. Figure 6 is divided into two regions, a feasible region in which all combinations of average yield and coefficient of variation can be obtained by some combining of slope and intercept parameters from equation 2.1, and the non-feasible region which is unobtainable using the harvest strategies described by equation 2.1 (and probably unobtainable by any harvest strategy).

The dashed boundary between the two regions is the most interesting line, since any point to the left of the line is inferior to some point on the line. That is, for any point inside the feasible region there exists a point on the boundary that has as good or better coefficients of variation and average yield.

Also shown in figure 6 are a series of circles that represent possible constant harvest rate policies. As we move from left to right, harvest rates increase. Note that the coefficient of variation changes very little, only the average yield increases. One fixed

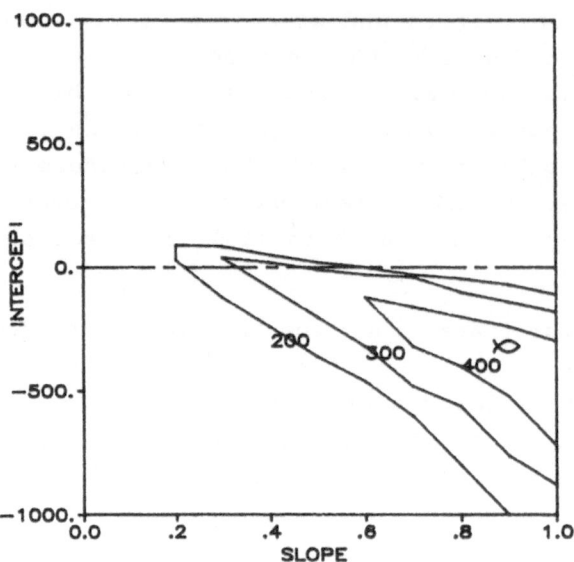

Figure 5. Average annual yield contours for different policies defined by a slope and an intercept for mixed stocks with different productivities and uncorrelated random variation.

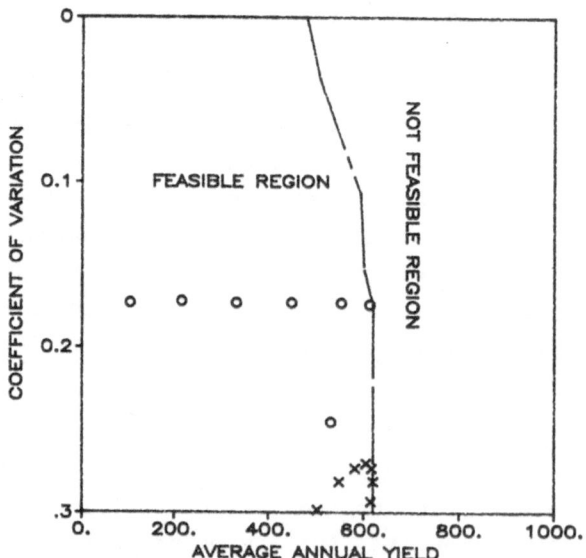

Figure 6. This graph shows the region of possible combinations of coefficient of variation and average annual yield that can be obtained by policies defined by equation 2.1 . Circles represent different constant harvest rate policies and X's represent different fixed escapement policies.

harvest rate policy lies right on the boundary of the feasible region, and is therefore preferred at some specified level of trade-off between yield and low variability. A similar series of X's represents possible fixed escapement strategies. The best fixed escapement strategy produces a yield roughly comparable to the best harvest rate strategy but with a much higher coefficient of variation. The top line of policies with a 0.0 coefficient of variation represents constant catch strategies.

## 5. Discussion

Although the results presented above are only numerical explorations of some specific assumptions about the nature of mixed stock fisheries, a number of interesting observations emerge. The most significant is the demonstration for a few cases that in managing mixed stock fisheries with uncorrelated errors, fixed escapement is not optimal. I have not been able to determine the true optimal policy due to the high dimensionality of the model, but I suspect some analysis or stochastic dynamic programming could find true optimal policies in such cases. The presence of ridges of roughly equal average catch is important as soon as managers begin to explicitly make trade-offs between average yield and variability of yield. We have seen that if a logarithmic utility is used to represent this trade-off, the optimum policy is very close to being a constant harvest rate strategy.

The vast majority of technical fisheries literature is based on the unit stock concept and we saw earlier that when this assumption is violated, fixed escapement policies are no longer optimal for maximization of average catch. Are most exploited stocks unit stocks or are they mixed? In Pacific salmon, where much of the theory of mixed stock fisheries was developed, nearly all stocks are explicitly recognized to be mixed. Different freshwater habitats are known to provide different survival rates and fish are known to be reasonably faithful to their natal streams. I know of no Pacific salmon fishery that could confidently be called a single stock fishery.

It is more difficult to determine the heterogeneity of other stocks, particularly in the major demersal and pelagic fisheries of the world. I would argue that the best default assumption is that they are mixed and the optimal fishing policy will therefore probably look much more like a fixed harvest rate policy than a fixed escapement policy. The stock concept international symposium contained many papers illustrating the discreteness of sub-stocks within stocks managed as unit stocks. Almost all of these studies concentrated on identifying discrete populations based on morphology or biochemistry. The mixed stock effects I have described above do not require genetic differentiation; the stocks could be genetically identical and different habitat quality alone could be responsible for varying productivity.

It is also clear that the second major issue, after stock discreteness, is correlation between stocks. I have examined only two extreme cases, totally correlated and totally uncorrelated. Examining the correlation structure of stocks fished together is obviously most important.

The numerical explorations presented here should be followed by more rigorous analytic work. Deriso's paper in this volume explores some of these problems, particularly optimal harvest policies with logarithmic objectives. Unfortunately I suspect that those models that are biologically realistic enough to represent stock structure will be quite difficult to analyze. This should not prevent managers from exploring alternative harvest strategies by simulation on models they consider reasonable.

## Acknowledgements

No thanks would be too much for Rick Deriso, Ralf Yorque and an anonomous barmaid in Kodiak, Alaska who all contributed to the formulation of the ideas in this paper.

## References

1. Ricker, W.E. (1958), Maximum sustained yields from fluctuating environments and mixed stocks. J. Fish. Res. Bd. Canada 15: 991-1006

2. Clark, C.W. (1976), Mathematical Bioeconomics. John Wiley, New York.

3. Reed, W.J. (1978), The steady state of a stochastic harvesting model. Math. Biosc. 41: 273-307

4. Reed, W.J. (1979), Optimal escapement levels in stochastic and deterministic harvesting models. J. Env. Econ. Mgmt. 6: 350-363

5. Reed, W.J. (1980), Optimum age-specific harvesting in a nonlinear population model. Biometrics 36: 579-593

6. Beddington, J.R. and D.B. Taylor (1973), Optimum age specific harvesting of a population. Biometrics 29: 801-809

7. Walters, C.J. and R. Hilborn (1976), Adaptive control of fishing systems. J. Fish. Res. Bd. Canada 33: 145-159

8. Walters, C.J. and R. Hilborn (1978), Ecological optimization and adaptive management. Ann. Rev. Ecol. Syst. 9: 157-188

9. Walters, C.J. (1981), Optimum escapements in the face of alternative recruitment hypotheses. Can. J. Fish. Aquat. Sci. 38: 678-689

10. Ludwig, D. and C.J. Walters (1982), Optimal harvesting with imprecise parameter estimates. Ecol. Modelling 14: 273-292

11. Smith, A.D.M. and C.J. Walters (1981), Adaptive management of stock-recruitment systems. Can. J. Fish. Aquat. Sci. 38: 690-703

12. Paulik, G.J., A.S. Hourston and P.A. Larkin. (1967), Exploitation of multiple stocks by a common fishery. J. Fish. Res. Bd. Canada 24: 2527-2537

13. Hilborn, R. (1976), Exploitation of multiple stocks by a common new fishery: a new methodology. J. Fish. Res. Bd. Canada 33: 1-5

14. Clark, C.W. and G.P. Kirkwood. (1984), On uncertain renewable resource stocks: optimal harvest policies and the value of stock surveys. The Institute of Applied Mathematics and Statistics, University of British Columbia, Vancouver B.C. Tech. Rept. 84-4

15. Hilborn, R. (1985), Expected changes in stock recruitment parameters when harvesting mixed stock fisheries. Can. J. Fish. Aquat. Sci: IN PRESS

# PATHOLOGICAL BEHAVIOR OF MANAGED POPULATIONS WHEN PRODUCTION RELATIONSHIPS ARE ASSESSED FROM NATURAL EXPERIMENTS

Carl J. Walters
Institute of Animal Resource Ecology
University of British Columbia
Vancouver, B.C.   V6T 1W5

## ABSTRACT

The assessment of fish production rates as a function of stock size is central to fisheries management, and this assessment obviously requires observations from a range of stock sizes.  When stock size variation is due mainly to natural, random disturbances that are treated as "natural experiments", the resulting sample may give a very distorted (biased) picture of the average production relationship. The bias will be especially bad when the natural disturbances occur in autocorrelated sequences over time.  These results suggest that there is no safe substitute for deliberate, controlled experimentation with stock size.

## INTRODUCTION

Decisions concerning optimum exploitation rates and/or escapements in managed populations are based on two types of information:  (1) prior expectations about productivity, based on past experience with related species and on calculations using biological observations of rate processes (growth reproduction, survival); and (2) direct estimates of stock response at different stock levels, obtained by analysis of historical monitoring data on the stock itself.  Estimates of type (1) presumably give way over time to estimates of type (2), since the type (1) estimates will inevitably contain qualitative errors of omission about processes that are locally important, along with quantitative errors due to small sample sizes and to processes that locally impact the quantitative levels of production parameters.

Few scientists would argue that there is ultimately any substitute for direct experience in quantifying the response patterns of each managed stock; the key issue is how to gain that experience in a safe and reliable fashion over time.  To quantify response patterns nessarily requires observations of production rates across a range of stock sizes, since density related changes in key rates (especially recruitment) define what exploitation regimes (decision choices) are sustainable by the stock.  Biological research may be essential to the

design of monitoring systems and the identification of key processes
to be monitored, but such research cannot experimentally provide the
range of stock sizes necessary to estimation of production relationships
in the field (i.e., for the stock as a whole). Here then is a fundamental point of controversy: should resource management attempt to
stabilize stock sizes and/or production rates at safe and reasonably
productive levels, or should management instead deliberately permit or
actively induce informative variation in stock sizes through variable
harvest regulations? This point of controversy is the basic research
issue of "adaptive management" (Walters and Hilborn 1976, 1978; Walters
1981, 1985; Ludwig and Walters 1982).

A vivid example of the controversy occurred recently in the State
of Washington (L. LeStelle, pers. comm.). In a court case there, the
Quinault Indian band argued for managing their chinook salmon using an
actively adaptive escapement policy, in which high escapements would be
deliberately allowed in years of large runs to test the possibility that
higher escapements (than in recent years) would result in higher
production rates. The State of Washington Department of Fisheries
opposed this policy, and argued instead for maintenance of escapement
at "standard" levels estimated to be optimum from analysis of spawning
grounds and juvenile rearing habitat; State biologists argued further
that their standards would inevitably be tested in any case, through
uncontrollable variation in escapement levels. The Indians won this
case, but a key point was not resolved: can we expect "natural
experiments" (uncontrollable variation in stock due to environmental
factors) to provide a sufficiently variable and unbiased sampling of
stock sizes and responses to make it unnecessary or unwise to introduce
further variation through deliberate management experiments?

In this paper I will use simple models to show that natural
experiments can lead to very misleading estimates of the relationship
between stock size and production rates. By natural experiments I will
mean situations where natural variation in production rates (environmental
effects on recruitment and survival) is permitted to cause stochastic
variation in stock size, and the stock sizes thus sampled are used to
assess the production relationship. I will show that such experiments
can (1) lead to bias in stock-recruitment parameter estimates, so that
production rates at low stock sizes are overestimated and the optimum
escapement is underestimated; (2) systematically favor incorrect
qualitative hypotheses (low or high) about the production relationship;
and (3) lead to the stock being systematically driven down to lower
levels than are actually optimum. These results apply even in the

simplest case, when unpredictable "environmental effects" are
uncorrelated over time; I briefly discuss the more realistic and
dangerous situation where unmodelled effects are persistent (autocorrelated) over time.

DEFINITION OF "NATURAL EXPERIMENTS" IN POPULATION MANAGEMENT

There are two sources of uncontrolled variation that a manager
might use to obtain informative variation in stock sizes without
resorting to deliberate experimentation with exploitation rates: (1)
variation in natural production rates due to "environmental" factors,
and (2) variation in exploitation rates due to unregulated changes in
harvesting effort (growth in fleet size, strikes that shut down the
harvesting operation, etc.). Here I shall consider only the first of
these sources; the second source will not cause estimation difficulties
unless the exploitation variations are correlated with production rate
variations.

As noted in Figure 1, alternative management responses to natural

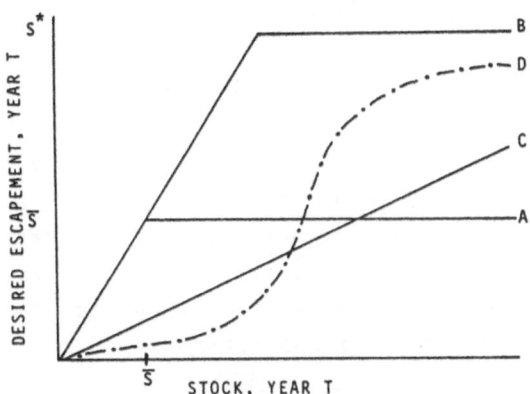

FIG. 1. Alternative harvesting policies.
A - fixed escapement at historial
average $\bar{S}$; B - probing policy
aimed at achieving experimental
escapement $S^*$; C - fixed exploitation rate policy; D - disruptive
policy that would tend to drive
the stock to extreme levels.

variation in production rates can be visualized as alternative feedback
policies for setting the escapement level in any year (stock after
harvest, or equivalently the exploitation rate) in relation to the

stock size resulting from variations previous to that year. Relative to the historical average escapement $\bar{s}$, the fixed escapement policy A in Figure 1 would eliminate (or minimize if stock is less than $\bar{s}$) informative variation. The "upward probing" policy B would allow informative increase in escapement relative to $\bar{s}$. The "fixed exploitation rate" policy C represents a simple and realistic compromise between these extremes; it permits all natural variations in stock size to influence the escapement (and future stock sizes) to a moderate degree.

In the following sections I will use policy C in Figure 1 as the prototype for natural experimentation. This policy involves neither the abrupt harvest changes of the probing policy B, nor the severe damping of variation implied by policy A. Obviously a variety of other functions (strategies) could be devised for responding to stock size so as to allow some informative variation; such policies might be less daring (i.e. compromise between A and C), but would otherwise be considered deliberately experimental (i.e., curve moving up from A to B would mean deliberately experimenting when the opportunity of a large stock size presents itself).

It is worth commenting here about a problem of definition, raised by why the fixed escapement policy A in Figure 1 is set so as to give an escapement equal to the historical average $\bar{s}$. To draw curve A with another breakpoint (i.e., like B) would be to assert that it is believed that the historical average policy $\bar{s}$ was not in fact optimum, and that this assertion is worth testing. In other words, any fixed escapement policy other than A as drawn would be an actively adaptive (deliberately experimental) policy, calling for deliberate departure from historical experience ($\bar{s}$) and hence having uncertain consequence. Likewise a strongly disruptive policy, such as curve D in Figure 1, would be seen by most managers as deliberately experimental.

BIAS IN PARAMETER ESTIMATION

Suppose now that the prototype, fixed exploitation policy is used to permit supposedly informative variation in stock sizes after harvesting each year. Suppose further that the exploitation rate is set at a level such that there are unlikely to be any major temporal trends (up or down) away from the historical average escapement $\bar{s}$. In this case the annual stock size prior to harvesting ("recruitment") will vary randomly due to environmental factors and due to variation in the escapement from the previous year; since a fixed proportion of the recruits are harvested, the escapement will also vary randomly.

The usual procedure for assessing the production relationship (recruitment as a function of escapement) in this situation would be to wait until several $S_t$, $N_{t+1}$ (escapement in year t, stock in year t+1) data pairs are available, then seek a functional relationship using various regression models and standard regression estimation procedures. In such analyses, $S_t$ is viewed as the "independent variable", while the "dependent variable" is either $N_{t+1}$ or some transformation of $S_t$, $N_{t+1}$ that is expected to have convenient statistical error properties. For example, if it is reasonable to assume that the functional relationship might be well approximated by Ricker's model $N_{t+1} = S_t e^{a-bS_t + W_t}$, where $W_t$ is a random environmental effect that is likely to be normally distributed (see Walters 1981, Peterman 1981), then an obvious transformation would be to take $y_t = \log(N_{t+1}/S_t)$. Then the production parameters a and b are estimated by linear regression, since $Y_t = a - bS_t + W_t$.

Suppose now that the estimation procedure can be formulated as a linear regression, to estimate a vector of production parameters $\underline{\beta}$ (e.g. a and b of Ricker model) from a vector of "outputs" $\underline{y}$ ($Y_i$ is the output for year i) and a "design matrix" X whose elements depend on the past states $S_t$, $S_{t-1}$, etc. For the Ricker model above, X is a 2 x n matrix, when n is the number of S-N data pairs available:

$$X = \begin{bmatrix} 1 & -S_1 \\ 1 & -S_2 \\ \vdots & \vdots \\ 1 & -S_t \end{bmatrix} \tag{1}$$

The column of 1's is associated with the a parameter, and the column of past escapements with the b parameter. In such situations, the usual regression estimator would be

$$\hat{\underline{\beta}} = (X'X)^{-1}X'\underline{y} \tag{2}$$

$\hat{\underline{\beta}}$ is maximum likelihood if the errors $W_t$ are normally distributed. Now, note that $\underline{y} = X\underline{\beta} + \underline{W}$ by assumption; substituting this relationship into equation (2) gives

$$\hat{\underline{\beta}} = (X'X)^{-1}X'(X\underline{\beta} + \underline{W}) \qquad (3)$$
$$= \underline{\beta} + (X'X)^{-1}X'\underline{W}$$

Equation (3) says that the regression estimates of β are in error ($\hat{\underline{\beta}} - \beta$) by the vector $(X'X)^{-1}X'\underline{W}$. Provided the X values are chosen by design to be independent of the process variations $\underline{W}$, this factor has expected value zero, i.e., the estimates $\hat{\underline{\beta}}$ are unbiased.

However, according to our supposition at the beginning of this section, the past states $S_t$ contained in the matrix X are indeed dependent on the process errors $W_{t-1}$, $W_{t-2}$, etc. That is, $S_t$ is a constant proportion of $N_t$ which in turn depends on $W_{t-1}$ and $S_{t-1}$, which in turn depends on $W_{t-2}$, and so forth. Consider the error vector $(X'X)^{-1}X'\underline{W}$; each element $W_i$ of $\underline{W}$ will have some (complex) effect on the subsequent (i+1, i+2, etc.) rows of X. In such situations there is no reason to expect or hope that $(X'X)^{-1}X'\underline{W}$ will have expected value zero, except as sample sizes (number of years data) become very large.

Thus the price of accepting natural experimentation (variation in S due to W alone) is, for realistically short time series, to accept bias in the parameter estimates. I have run a large number of Monte Carlo trials for the Ricker recruitment example (Walters, 1985a), to estimate the magnitude of this bias when samples of 10 stock-recruitment pairs are available, using Ricker a and b parameters and variances of W that are typical for Pacific salmon. The results of these trials were most disturbing: the a parameter, which determines the optimum exploitation rate, was typically overestimated by 30-50%; the b parameter was similarly overestimated, and the optimum escapement was underestimated by around 30%.

It is possible to develop correction procedures for the bias, but such procedures lead to estimated production relationships that in some sense do not "fit the facts" as well as the original biased estimates. In other words, the correction procedures lead to different hypotheses about the production relationship, but these hypotheses are based on strong assumptions about the statistical error relationships and may thus be no more credible than hypotheses based on standard estimation procedures. Instead of resolving uncertainties, natural experiments may simply lead to further confusion about how to interpret the data.

One way to avoid the bias implied by equation (3) would be to manage escapements so as to allow random variation in $S_t$, while somehow insuring that this variation is uncorrelated with $N_t$ (and thus $W_{t-1}$). The process of inseason management (openings and closures, effort limits, etc.) is likely to introduce some random variation in

escapements, due to factors such as monitoring errors and variation
in catchability coefficients.  However, such factors are unlikely to
introduce enough "natural" variation to destroy the $N_t$-$S_t$ correlation.

## MISLEADING TRENDS AND TRAPS AT LOW STOCK SIZE

The concerns raised in the previous section may seem highly
theoretical or distant from practical debate about management, since
they arise through the rather obscure mathematical machinery of
quantitative parameter estimation.  This section looks at a more
familiar management situation, namely where debate about the production
relationship has crystallized into two opposing viewpoints (alternative
hypotheses) that appear equally consistent with historical experience
(Figure 2).  In this common situation, stock size data are available

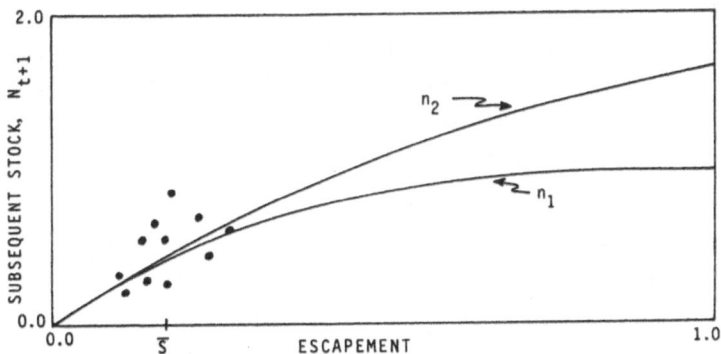

FIG. 2.  Two alternative hypotheses about the effect of
allowing higher escapements.

only for recent years, though the stock has been harvested for longer
and may be far below its natural equilibrium or "carrying capacity"
level.  Proponents of model $\eta_1$ in Figure 2 argue that the stock is not
currently overexploited, so there would be no benefit from reducing
harvest rates to allow more escapement.  Proponents of model $\eta_2$ contend
that the stock was depleted before monitoring even began, and that
harvest rates should be reduced so that more productive escapement
levels will be achieved for the future.

A compromise management policy in this situation would be to fix
the exploitation rate at a moderate level, so that some extra escapement
is allowed in the good years as a test of the $\eta_2$ hypothesis of Figure
2, yet harvests are not shut down in average years just to accomplish

such tests. In the following paragraphs I will mainly present
arguments and calculations based on such a fixed exploitation rate
policy. To add realism, I will also make a few comments about the
impacts of combining this policy with a weak process of "passive
adaptation" (Walters and Hilborn 1978), in which the data are
periodically reanalysed so as to reset exploitation rate targets based
on whichever hypothesis appears most likely to be correct. As we shall
see, passive adaptation can either mitigate or exaggerate the biases
associated with a fixed exploitation policy.

Assuming that one or the other hypothesis in Figure 2 is correct,
and not something in between or more extreme than either, Bayes theorem
can be used to calculate a simple measure (the Bayes posterior
probability) of credibility for each hypothesis over time. Suppose
that before examining any of the data, we assign the two hypotheses
equal prior credibilities, $P_o(\eta_1) = P_o(\eta_2) = 0.5$. Suppose further that
for any data point ($S_t$, $N_{t+1}$ pair) that is obtained, we can agree
about how to calculate the likelihoods $L(Y_t|\eta_i)$, or simply $L_{it}$, of
obtaining that $t\underline{th}$ observation given that model $\eta_i$ is true, for each
model i. Then according to Bayes theorem, the credibility of model i
after getting the $t\underline{th}$ data point, $P_t(\eta_i)$, is just

$$P_t(\eta_i) = \frac{L_{it}P_{t-1}(\eta_i)}{\sum_j L_{jt}P_{t-1}(\eta_j)} \tag{4}$$

(Here the denominator is the "probability of the data", and the sum -
j is over just the two models).

Equation (4) can be used to set up a very simple microcomputer
demonstration of how the credibility of each alternative hypothesis
is likely to evolve over time in relation to the escapement ($S_t$) time
series. In the demonstration calculations described below, I took
models $\eta_1$ and $\eta_2$ to both be Ricker curves, i.e. $\eta_1$: $N_{t+1} = S_t e^{1-S_t+W_t}$
and $\eta_2$: $N_{t+1} = S_t e^{1-0.5S_t+W_t}$, with normally distributed, mean zero
environmental effects $W_t$ in the exponents. In this case, the $L_{it}$ are
given simply by the normal density function:

$$L_{it} = \frac{1}{\sqrt{2\pi}\ \sigma_w} e^{-(\hat{W}_{it}^2/2\sigma_w^2)}$$

where $\sigma_w^2$ is the variance of environmental effects and $\hat{W}_{it}$ is the
estimated deviation for year t if model i is true (for example

$W_{1t} = \log(N_{t+1}/S_t) - 1 + S_t)$. I made the rather strong assumption that $\sigma_w^2$ is known in advance; in practice the assumed value for this parameter does not have a major effect on the equation (4) estimates, unless $\sigma_w^2$ is set very low or so high that both models are always assigned nearly equal $L_{it}$. For the tests, I used $\sigma_w = .05$.

Figure 3 shows a sample output from the demonstration program.

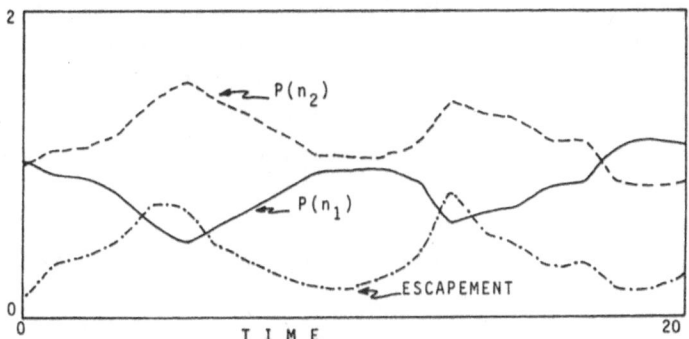

FIG. 3. Sample simulation of learning about which hypothesis in Fig. 2 is correct, when $n_1$ is in fact correct and exploitation rate is held constant.

In this Monte Carlo trial, the correct model was $n_1$, and the annual exploitation rate was set at 0.5. Notice that the credibility of $n_1$ remains near 0.5 (both models equally credible) until a natural experiment (accidental series of positive $W_{it}$ values) causes the spawning stock to increase. At high spawning stocks the models predict quite different responses ($\hat{W}_{it}$ likely to be quite different), and the correct model quickly achieves high credibility. The outcome of such natural experiments is easily understood, but note that the probability of these "lucky" outcomes is low: much yield might be lost while waiting for one of these experiments and managing according to the wrong model during the waiting period.

A disturbing pattern is evident in Figure 3 for the years prior to occurrence of a definitive natural experiment. Whenever relatively good (small positive deviation) years occur, there is a double effect: escapements increase toward more informative levels, but the positive deviations simultaneously favor model $n_2$ (i.e., the deviation from $n_2$ is smaller, so $L_{2t}$ is higher in equation 4). In other words, there is no way for an informative natural experiment to be initiated, without the initiating random event itself favoring one or the other hypothesis, unless by luck the event occurs at a starting stock size $S_t$ for which the alternative models happen to predict exactly the same outcome (i.e., assign the event the same probability of occurrence). As shown in Figure 3, this biasing effect (assignment of higher $P_t$ to model $\eta_2$) can be persistent over time so that the incorrect model is progressively favored.

If there is a weak adaptive response to the shifting probabilities, so that escapements are deliberately increased further after a few natural events lead to higher probabilities for $\eta_2$ (along with moderate stock increase), then this adaptive response will hasten the discovery that the correct model was in fact $\eta_1$. That is, deliberate escapement increases will move $S_t$ to levels where the predictions of $\eta_1$ and $\eta_2$ are more distinct, so the odds of misleading deviations are reduced; passive adaptation will then work against the bias described in the previous paragraph.

Figure 4 shows an even more disturbing sample trial. In this

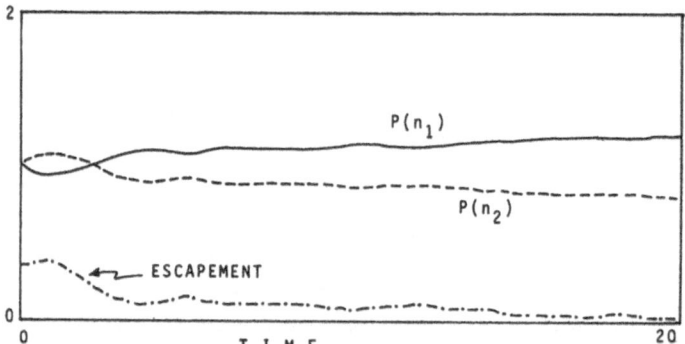

FIG. 4. Sample simulation of probabilities assigned to the models in Fig. 2; $\eta_2$ is in fact correct and the stock size is declining due to a high fixed exploitation rate.

trial, the correct model was changed to $\eta_2$ and the annual exploitation rate was set at 0.6 rather than 0.5 so as to produce a slow decline in stock size. Notice in this case that the credibility assigned to model $\eta_1$ increases slowly over time, i.e. the wrong model is systematically favored. The explanation for this phenomenon is simple enough: whenever a substantial negative random event (large negative $W_t$) occurs, this event has the double effect of favoring model $\eta_1$ and simultaneously driving the stock size $S_{t+1}$ down. The higher exploitation rate tends to prevent recovery from the lower $S_{t+1}$ level, and the two models make less distinctive predictions at this level so that the shift in credibilities at time t is not fully cancelled even if a similarly strong positive event (positive $W_{t+1}$) occurs at time t+1.

Figure 4 is a simple scenario for a pathological process of passive adaptation. Suppose that whenever negative deviations $W_t$ occur, reducing stock size and favoring model $\eta_1$, there is a tendency for the management agency to treat these "false signals" as representative responses and to act accordingly by preventing escapement levels $S_{t+1}$, $S_{t+2}$ ... from increasing even if positive deviations occur later (The harvest rate 0.6 rather than 0.5 was chosen to represent such a decision process). The effect of such a decision sequence is to progressively lock the system into more uninformative (lower $S_t$) states, at least with regard to the formal calculation of probabilities for the alternative models. In this case passive adaptation acts not to reduce the biases due to natural escapements, but instead to exaggerate them. Hilborn (1979), Smith (1979) and Ludwig and Hilborn (1983) have noted a similar effect in simulations of performance for passive adaptive control of surplus production systems; there is a tendency for the controller to become "locked in" on low estimates of the optimum stock size, and to never allow low enough harvest rates for the stock to increase enough for formal estimation procedures to reveal that it was overexploited in the first place.

One might argue that the situation shown in Figure 2, and scenarios derived from it, is a special configuration contrived to demonstrate the worst possible outcomes of natural experiments. To show that this is not the case, consider the alternative qualitative situation in Figure 5. In this situation, the stock has historically not been

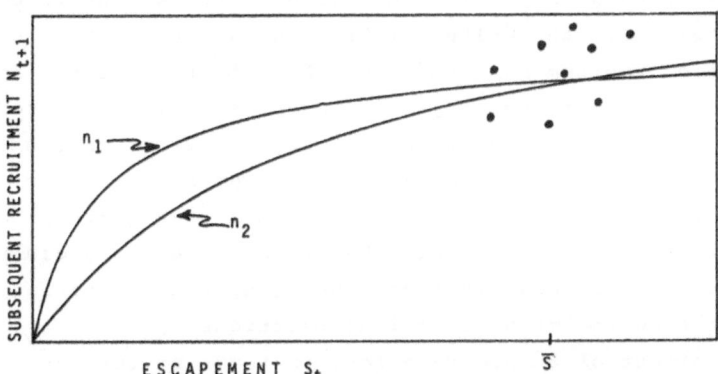

FIG. 5. Two alternative hypotheses about the effect of reducing the escapement level allowed.

heavily exploited, and there are two general hypotheses about the impact on production of allowing higher harvest rates. Suppose it is decided to wait for a natural experiment (poor survival year, negative $W_t$) to reduce stock sizes to an informative level. The occurrence of this event (or sequence of events) will immediately favor model $\eta_2$ even if $\eta_1$ is correct. Then if the adaptive response to this misleading evidence is to reduce harvests so as to rebuild $S_t$, few informative observations will be gathered; if model $\eta_1$ is in fact correct, an opportunity for increased yields will have been lost.

AUTOCORRELATION IN DEVIATIONS FROM MODEL PREDICTIONS

In both the general situations shown in Figures 2 and 5, the discussion so far has assumed that there are no temporal drifts or cyclic changes in the "parameters" defining the mean production relationship. Such changes would be statistically evident as (i.e. would appear to involve) autocorrelation in the deviations $W_t$ over time, and would reflect the impacts of biological and physical interactions that are hidden in the aggregate production relationship between stock and recruitment. Indeed, the use of simple models with autocorrelation in deviations may be the best strategy currently available for dealing with ecological complexity; the track record has not been good for the strategy of building more complex models that explicitly represent various interactions. In any case, autocorrelation in errors represents a third general situation that, unlike Figures 2 and 5, cannot easily be represented in terms of simple graphical alternatives.

Little is known about optimum feedback policies for situations involving autocorrelation in environment effects. When the mean production relationship is poorly understood, with parameters changing slowly over time, Smith and Walters (1981) and Walters (1985) have presented dynamic programming results which indicate that the optimum policy should involve periods of passive adaptation interspersed with episodes of strong probing to obtain regular fixes on the drifting parameter values. When the parameters may change rapidly (low auto-correlation in prediction errors), some preliminary dynamic programming results (Walters, in prep.) indicate that there is a threshold rate of change above which it is best to treat the changes as a stationary noise process around the mean historical relationship.

The basic effect of autocorrelation, in terms of the concerns raised in previous sections, is to exaggerate any biases or misleading probabilistic assessments based on natural experiments. A current example of possible confusion arising from temporal correlation (in environmental effects on recruitment) is with chinook and coho salmon. Over most of the Pacific coast, wild (i.e., natural, nonhatchery) escapements of these species have been declining for at least the last 15-20 years, while hatchery stocks have been increasing. A peak in catch occurred in the mid 1970's, and catches are now (1984) declining. This decline has been widely attributed to overexploitation of wild stocks (i.e., following a curve like $\eta_2$ in Figure 5). But oceanographic conditions have been unusual over the last 10 years, with high surface temperatures and a dramatic El Niño event. Are we seeing overexploitation, or a natural climate "cycle" that will correct itself? Another obvious example of this confusion is the famous "Thompson-Burkenroad debate" about impacts of fishing versus environment on the stocks of Pacific halibut (Skud, 1975).

In a recent assessment of stock recruitment relations for B.C. salmon, F. Wong (1982) found strong ($\geq 0.4$) autocorrelations at lags of 1 to 3 years in deviations from the recruitment curves for 25% of the 47 aggregate stocks (species in large areas) that he examined. Most of the high correlations were for areas where several stocks are harvested together, and may thus just reflect errors in allocating mixed stock catches to the rivers of origin. Nevertheless there are several distinctive cases, including the major Fraser River sockeye stock (Figure 6), where the striking autocorrelation is apparently not due to such errors. The Fraser sockeye case has been repeatedly cited (e.g. Walters and Hilborn 1976, Walters 1985b) as an example of the situation in Figure 2. Yet the autocorrelation structure in Figure 6 implies that an experiment (natural or deliberate) involving increased escapement could give misleading results unless the high escapements are maintained for at least 10-15 years; on shorter time scales, "runs" of positive or negative $W_t$ values could easily favor a recruitment hypothesis that is not representative of long term average response.

The scenarios discussed above do not demonstrate that natural experiments are always dangerous or misleading. Rather, these scenarios show that there is a double effect of each natural variation (immediately favor one extreme hypothesis, send the state to a more or less informative level), with longer term consequences that depend on luck and on the direction of management response to new states of affairs. There is a significant long term danger only in those situations where the management response (or nominal exploitation pattern) based on the "best fitting model" or "best available estimate" would send the system into a less informative range of states.

102

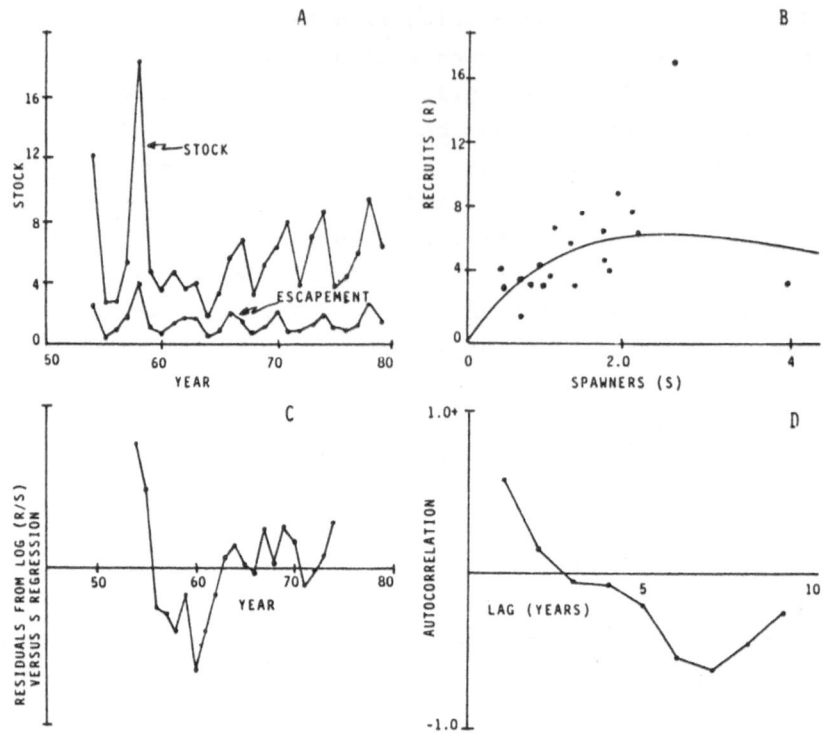

FIGURE 6.  Stock and recruitment statistics for Fraser River sockeye, from F. Wong
(1982).   A - time series of stock sizes and escapements; B - relationship
between escapement in year t and resulting stock in year t+4; C - time
series of deviations from stock-recruitment relation in panel B, measured
as deviations from regression of log ($N_{t+4}/S_t$) on $S_t$; D - autocorrelation
function for deviations from panel C (only the tag 1 autocorrelation is
statistically significant).

Given reasonably long historical time series, it is possible to
correct for obvious biases due to autocorrelation effects (i.e., model
the autocorrelation structure explicitly, and assign a likelihood to
each new observation $W_t$ that reflects this structure), and to the
double effect of each informative deviation (i.e., omit data from
such years when doing probability calculations, or use Monte Carlo
procedures to estimate the likely magnitude of bias).   However, the
development of correction procedures involves making some strong
assumptions about the $W_t$ stochastic process, and these assumptions
may be as badly misleading as the uncorrected assessments.

The fundamental problem is that in analysing the production time
series for any one population, there is no control (in the scientific
sense) situation against which to measure the effect of $W_t$ separately
from the effect of changing stock size. This way of stating the
problem leads immediately to a question that has received little
attention in fisheries stock assessment:  when there are several stocks
of the same species in a region (a very common situation), can these
stocks be used as controls for one another?  In other words, do the
stocks consistently exhibit "shared" environmental effects $W_t$, so that
a "pooled" (across stocks) estimate of each $W_t$ can be obtained and
used to filter the data for each stock?  These questions deserve
further research, but the discovery that shared effects are large
would not provide provide a justification for relying on natural
experiments:  given large shared effects, the stocks should also
show highly correlated variations in stock size (unless they are
deliberately managed differently, which would not be a natural experiment)
which in turn would make estimation of the shared environmental effects
(as opposed to stock size effects) more difficult and potentially
misleading.  However, it may be much easier to politically and
economically justify deliberate experimentation with harvest policies,
when the direct costs and risks of such experiments are shared across
several stocks.

CONCLUSION

Analysts of fisheries time series data have had to rely mainly on
variations caused by environmental factors and uncontrolled effort
development to provide contrasting stock sizes.  In such cases,
standard statistical procedures may give very misleading estimates of
the relationship between production rates and stock size even if the
stock sizes are measured without error.  It is not clear that the
difficulty can be resolved through development of procedures for bias
correction.  Since natural experiments do not provide a reliable basis
for acquiring better estimates of production relationships, we need
to look more carefully at the question of whether it is worthwhile to
include deliberate experimentation as part of harvest management
programs.

# REFERENCES

Hilborn, R.   1979.   A comparison of alternative fisheries control systems that use catch and effort data.   J. Fish. Res. Bd. Canada, 36: 1477-1489.

Ludwig, D. and R. Hilborn.   1983.   Adaptive probing strategies for age structured fish stocks.   Can. J. Fish. Aquat. Sci., 40: 559-569.

Ludwig, D. and C.J. Walters.   1982.   Optimal harvesting with imprecise parameter estimates.   Ecological Modelling, 14: 273-292.

Peterman, R.M.   1981.   Form of random variation in salmon smolt-to-adult relations and its influence on production estimates.   Can. J. Fish. Aquat. Sci., 38: 1113-1119.

Skud, B.E.   1975.   Revised estimates of halibut abundance and the Thompson-Burkenroad debate.   Int. Pac. Halibut Comm., Sci. Rept. #56, 35 pp.

Smith, A.D.M.   1979.   Adaptive management of renewable resources with uncertain dynamics.   Ph.D. thesis, Univ. of B.C., 192 pp.

Smith, A.D.M. and C.J. Walters.   1981.   Adaptive management of stock-recruitment systems.   Can. J. Fish. Aquat. Sci., 38(6): 690-703.

Walters, C.J.   1981.   Optimum escapements in the face of alternative recruitment hypotheses.   Can. J. Fish. Aquat. Sci., 38(6): 678-689.

Walters, C.J.   1985a.   Bias in the estimation of functional relationships from time series data.   Can. J. Fish. Aquat. Sci., 42(1): 147-149.

Walters, C.J.   1985b.   Adaptive  Policy Design in Renewal Resource Management.   Accepted for publication by MacMillan Pub. Co., Inc., N.Y., approximately 300 pp.

Walters, C.J. and R. Hilborn.   1976.   Adaptive control of fishing systems.   J. Fish. Res. Bd. Canada, 33: 145-159.

Walters, C.J. and R. Hilborn.   1978.   Ecological optimization and adaptive management.   Annual Review of Ecology and Systematics, 9: 157-188.

Wong, F.   1982.   Analysis of stock recruitment dynamics of B.C. salmon. MS thesis, Univ. of B.C.   221 pp.

# SEARCH MODELS IN FISHERIES AND AGRICULTURE

Marc Mangel[*]
Department of Mathematics
University of California
Davis, CA. 95616

## Abstract

Various stochastic models associated with search problems in fisheries and agriculture are developed. This includes a discussion of binomial, Poisson, and negative
binomial distributions and an introduction to diffusion processes. Random search
models are discussed, followed by discussion of exhaustive search models. Two applications are presented. The first is a study of the relationship between catch per
unit effort and stock abundance in fisheries. The second is a problem in trap spacing for the detection of an invading agricultural pest.

> This paper is dedicated to the memory
> of Steve Goodman. His music
> will be missed by many
> of us.

## 1. Introduction

Search problems arise in numerous contexts in resource management. A few examples are the following. Schools of fish must be found before they can be harvested (1,2,3). Foraging is essentially a search process (4,5). Scouting for pests in
agricultural pest control is a search problem (6,7,8). These search problems can be
broadly classified as descriptive or prescriptive. When a descriptive search problem
is solved, one ends up with a method for describing the time evolution of probability
densities and similar quantities. The solution of a prescriptive problem tells one
how to search over time and/or space. This paper is concerned mainly with descriptive search problems and the underlying mathematical models. Some prescriptive
problems are treated in (3,5,6,7).

Section 2 contains a description of the basic statistical models used in search
problems. These include the binomial, Poisson, and negative binomial distributions,
a new extension of the negative binomial distribution, transformations to normality,
and diffusion models. Section 3 is concerned with _random_ search. A theory, due to
Koopman (9), is extended by application of renewal theory. Section 4 contains a

---

[*]Also: Departments of Agricultural Economics and Entomology, Aquaculture and
Fisheries Program.

summary and exposition of Neyman's work (10) on exhaustive search. Sections 5 and 6 are concerned with specific examples: Section 5 on the relationship between catch per unit effort (CPUE) and abundance in fisheries and Section 6 on trap spacing in agricultural pest control.

## 2. Basic Search Models and Extensions

This section contains a sequence of models, with increasing complexity, that are useful in understanding and solving search problems.

### 2.1. Binomial Distribution

Imagine a large region $\Omega$ of area $A$ that contains $N$ objects. Suppose that the objects are randomly distributed in $\Omega$ and that area is searched at a rate $\alpha t$ where $t$ is search time and $\alpha$ is area searched per unit time. Assume that an object is only detected once, if at all. For example, if the object is a school of fish it is removed upon harvest or somehow tagged so that it won't be detected again. Suppose that $X(t)$ is the number of objects found in search time $t$. Then the probability of detecting a single object is $\alpha t/A$. Assuming that the $N$ objects are randomly distributed and that $\alpha t/A < 1$ and detections of different objects are independent events, then (11,12)

$$\Pr\{X(t) = k\} = \binom{N}{k}\left(\frac{\alpha t}{A}\right)^k\left(1 - \frac{\alpha t}{A}\right)^{N-k} \tag{2.1}$$

Equation (2.1) is a binomial distribution with parameters $N$ and $p = \alpha t/A$. Thus

$$\left.\begin{array}{l} E\{X(t)\} = \dfrac{N\alpha t}{A} \\[2ex] \text{Variance}\{X(t)\} = N\left(\dfrac{\alpha t}{A}\right)\left(1 - \dfrac{\alpha t}{A}\right) \end{array}\right\} \tag{2.2}$$

The binomial distribution is basic when randomness is involved and depletion (removal) is taken into account. Under certain circumstances, simplifications arise.

### 2.2. Poisson Distribution

Suppose now that $N$ and $A$ go to infinity in such a way that $\alpha t/A \to 0$ but $N\alpha t/A \to \lambda t$ where $0 < \lambda < \infty$. This limit corresponds to the Poisson approximation to the binomial distribution (11,12) which is

$$\Pr\{X(t) = k\} = \frac{e^{-\lambda t}(\lambda t)^k}{k!} \, . \tag{2.3}$$

Equation (2.3) is often derived in another fashion, which turns out to be helpful for extensions. This approach is based on the following assumption.

$$\left.\begin{array}{l} \Pr\{\text{another detection in } (t,t+dt)\} = \lambda \, dt + o(dt) \\[2mm] \Pr\{\text{no detection in } (t,t+dt)\} = 1 - \lambda \, dt + o(dt) \end{array}\right\} \tag{2.4}$$

This equation by itself is the starting point for the derivation of the Poisson distribution, so $\lambda$ need not have the interpretation $\lambda = N\alpha/A$.

If $p_k(t) = \Pr\{X(t) = k\}$, then (2.4) leads to

$$\left.\begin{array}{l} p_0(t+dt) = p_0(t)(1-\lambda \, dt) + o(dt) \\[2mm] p_k(t+dt) = p_k(t)(1-\lambda \, dt) + p_{k-1}(t)\lambda \, dt + o(dt) \end{array}\right\} \tag{2.5}$$

$$k > 1$$

Equation (2.5) is converted to the following differential equations

$$\left.\begin{array}{l} \dfrac{dp_0}{dt} = -\lambda p_0(t) \\[4mm] \dfrac{dp_k}{dt} = -\lambda p_k(t) + \lambda p_{k-1}(t) \end{array}\right\} \tag{2.6}$$

The initial conditions $p_0(0) = 1$, $p_k(0) = 0$, $k \geq 1$, lead to (2.3).

For the Poisson distribution

$$E\{X(t)\} = \mathrm{Var}\{X(t)\} = \lambda t. \tag{2.7}$$

Very often in ecological and resource sampling, which takes place over space as well as time, it is found that the variance of the number of objects discovered far exceeds the mean. This indicates that the Poisson distribution is not sufficient as a model; it must be improved upon.

## 2.3. Negative Binomial Distribution

Imagine now that the Poisson model (2.3) or (2.4) holds <u>locally</u> at points in $\alpha$, but that $\lambda$ varies on a global scale in $\alpha$. Assume that the precise variation of $\lambda$ over $\alpha$ is not known. This means that at each point $\lambda$ has a distribution. The distribution chosen here is a gamma distribution with parameters $\nu$ and $\alpha$:

$$\Pr\{\lambda \in (y,y+dy)\} = \frac{e^{-\alpha y} y^{\nu-1} \alpha^\nu}{\Gamma(\nu)} \, dy \tag{2.8}$$

The mean of $\lambda$ is $\nu/\alpha$.

In this case, (2.3) is interpreted as a conditional distribution of X(t), given $\lambda$. The unconditional distribution is found by integrating (2.3) against (2.8):

$$\Pr\{X(t) = k\} = \int_0^\infty \frac{e^{-\lambda t}(\lambda t)^k}{k!} \frac{e^{-\alpha\lambda}\lambda^{\nu-1}\alpha^\nu}{\Gamma(\nu)}\, d\lambda$$

$$= \frac{\Gamma(k+\nu)}{\Gamma(\nu)} \frac{t^k}{k!} \frac{\alpha^\nu}{(\alpha+t)^{k+\nu}} \tag{2.9}$$

This is a negative binomial distribution (11). It can be put into a more recognizable form by setting

$$p = \frac{\alpha}{\alpha+t} \tag{2.10}$$

When this is done, (2.9) becomes

$$\Pr\{X(t) = k\} = \binom{k+\nu-1}{k} p^\nu (1-p)^k \tag{2.11}$$

The mean and variance of X(t) are now given by

$$\left.\begin{aligned} E\{X(t)\} &= \frac{\nu q}{p} = \frac{\nu}{\alpha} t \equiv m(t) \\ \mathrm{Var}\{X(t)\} &= m(t) + \frac{m(t)^2}{\nu} \end{aligned}\right\} \tag{2.12}$$

Equation (2.12) shows that $\mathrm{Var}\{X(t)\} \geq E\{X(t)\}$ and equality holds only when $\nu \to \infty$. (It is easily seen that if $\nu$, $\alpha \to \infty$ in (2.8) in such a way that $\nu/\alpha = \bar{\lambda}$ is constant, then the density (2.8) approaches a delta function centered at $\bar{\lambda}$.) Thus, to some extent the negative binomial distribution models the "overdispersion" or "patchiness" alluded to before. This behavior is nicely exhibited by a study of the respective coefficients of variation, CVs. The CV is defined by

$$\mathrm{CV} = \frac{\sqrt{\mathrm{Variance}}}{\mathrm{Expectation}} \tag{2.13}$$

so that the coefficient of variation for the Poisson distribution is

$$\mathrm{CV}_P = \frac{1}{\sqrt{\lambda t}} \tag{2.14}$$

and the coefficient of variation for the negative binomial distribution is

$$\mathrm{CV}_{NB} = \sqrt{\frac{1}{m(t)} + \frac{1}{\nu}} \; . \tag{2.15}$$

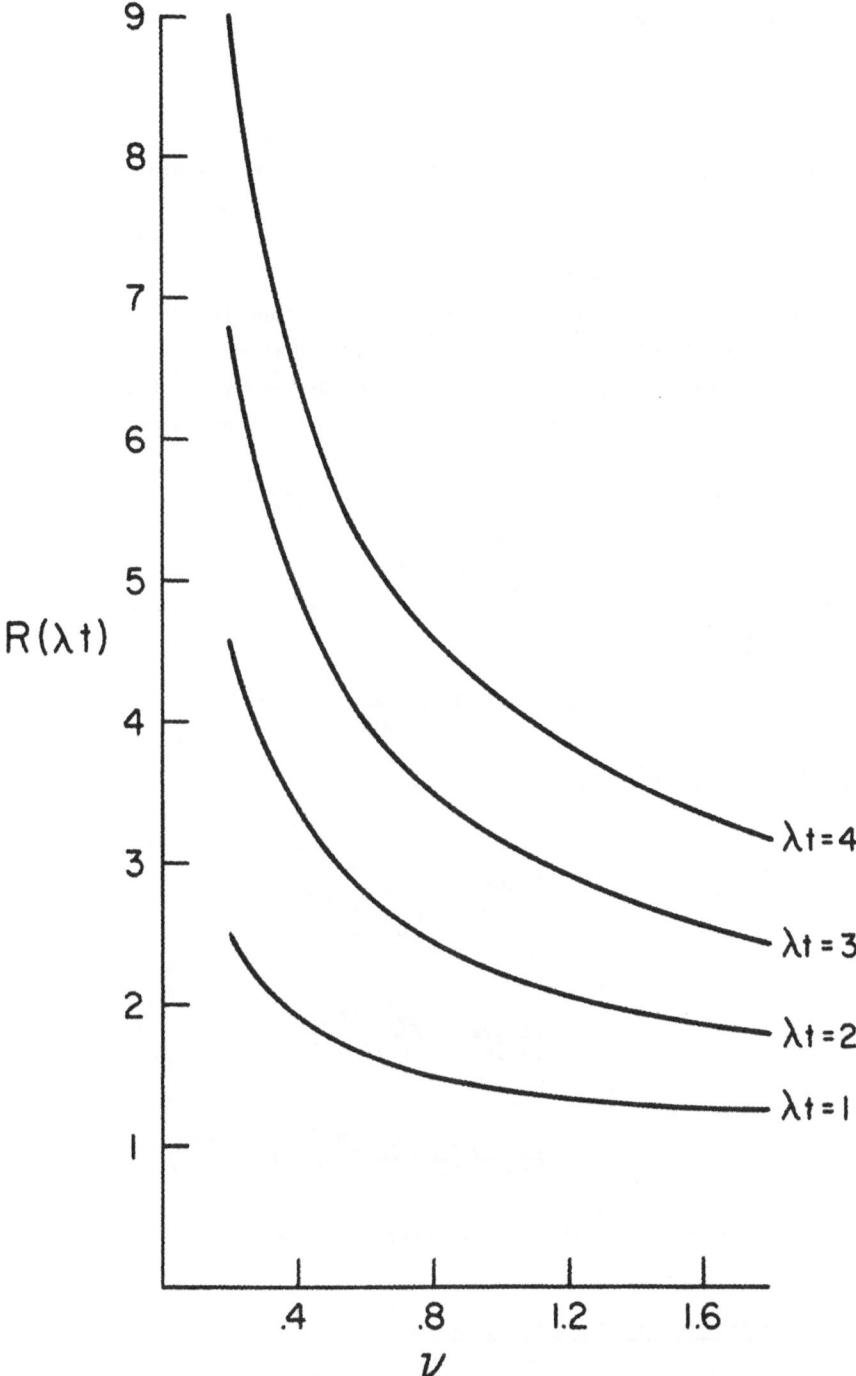

Fig. 2.1   Comparison of Coefficients of Variation of the
Poisson and Negative Binomial Distributions.

Figure 2.1 shows the ratio

$$R(\lambda t) = \frac{CV_{NB}}{CV_P}$$

for a number of values of $\lambda t$ and as a function of $\nu$. This figure clearly shows the overdispersion of the negative binomial distribution relative to the Poisson. The next question is then: can one do the same thing when depletion is important? This question is answered in the next section.

Observe from (2.12) that if $(\frac{\nu}{\alpha})t = m(t)$ is fixed, then the "overdispersion" is characterized by the parameter $\nu$. It is often argued that $\nu$ is a species characteristic, so that previous sampling experience can be used to estimate $\nu$.

Another commonly used form of the negative binomial distribution is

$$\Pr\{X(t) = n\} = \frac{\Gamma(n+k)}{\Gamma(k)n!}\left(1 + \frac{m}{k}\right)^{-k}\left(\frac{m}{m+k}\right)^{n}$$

where $k$ and $m$ are the parameters. Here $m$ is the sample mean and $k$ is the overdispersion parameter $(\text{Var}\{X(t)\} = m + m^2/k)$. This form can be related to (2.9) by rewriting (2.9) as follows

$$\Pr\{X(t) = n\} = \frac{\Gamma(n+\nu)}{\Gamma(\nu)n!}\left(\frac{t}{\alpha+t}\right)^{n}\left(\frac{\alpha}{\alpha+t}\right)^{\nu}.$$

Now set $m = \frac{\nu}{\alpha}t$ so that this equation becomes

$$\Pr\{X(t) = n\} = \frac{\Gamma(n+\nu)}{\Gamma(\nu)n!}\left(\frac{\frac{\nu}{\alpha}t}{\nu + \frac{\nu}{\alpha}t}\right)^{n}\left(1 + \frac{t}{\alpha}\right)^{-\nu}$$

$$= \frac{\Gamma(n+\nu)}{\Gamma(\nu)n!}\left(\frac{m(t)}{\nu + m(t)}\right)^{n}\left(1 + \frac{\frac{\nu}{\alpha}t}{\nu}\right)^{-\nu}$$

$$= \frac{\Gamma(n+\nu)}{\Gamma(\nu)n!}\left(\frac{m(t)}{\nu + m(t)}\right)^{n}\left(1 + \frac{m(t)}{\nu}\right)^{-\nu}.$$

Thus, $k$ and $\nu$ have exactly the same interpretations.


## 2.4. Extended Negative Binomial Distribution

Now consider an extension of the Poisson process to include depletion. In particular, replace assumption (2.4) by

Pr{another detection in $(t, t+dt) | n$ detections up to $t, \lambda$}

    $= (\lambda - n\epsilon)dt + o(dt)$ if $n\epsilon < \lambda$

    $= 0$             otherwise,

Pr{no detection in $(t, t+dt) | n$ detections up to $t, \lambda$}

    $= 1 - (\lambda - n\epsilon)dt + o(dt)$ if $n\epsilon < \lambda$

    $= 1$             otherwise.

$$(2.16)$$

These infinitesimal assumptions lead to a binomial distribution (3,11) in which

$$\text{Pr}\{n \text{ encounters in } (0,t) | \lambda\}$$

$$= \binom{\lambda/\epsilon}{n}(1 - e^{-\epsilon t})^n (e^{-\epsilon t})^{(\lambda/\epsilon)-n} \tag{2.17}$$

$$n = 0,1,2,\ldots,\lambda/\epsilon.$$

Observe that as $\epsilon \to 0$, $N \equiv \lambda/\epsilon \to \infty$ and $p \equiv 1 - e^{-\epsilon t} \to 0$ but $Np \to \lambda$ and (2.17) reduces to the Poisson distribution.

The negative binomial distribution is obtained by averaging a Poisson distribution with parameter $\lambda$ over a gamma distribution. In order to obtain the extended negative binomial (ENB) distribution, (2.17) will be averaged over a distribution on $\lambda$ (see also (3)).

In order to do this, define $(\lambda/\epsilon)_n$ by (13)

$$(\lambda/\epsilon)_n = (\frac{\lambda}{\epsilon} - n + 1)(\frac{\lambda}{\epsilon} - n + 2) \cdots (\lambda/\epsilon) \qquad \lambda/\epsilon \geq n$$

and set

$$(\lambda/\epsilon)_n = \sum_{j=1}^{n} A(j,n)(\lambda/\epsilon)^j. \tag{2.18}$$

Since $(\lambda/\epsilon)$ satisfies the recurrence $(\lambda/\epsilon)_{n+1} = (\frac{\lambda}{\epsilon} - n)(\lambda/\epsilon)_n$, it can be shown that the $A(j,n)$ in (2.18) satisfy the following recurrences

$$\begin{array}{l} A(n,n) = 1 \\ A(k,n+1) = A(k-1,n) - nA(k-1,n) \\ A(1,n+1) = -nA(1,n). \end{array} \tag{2.19}$$

Using this notation, (2.17) can be rewritten as

$$\Pr\{n \text{ encounters in } (0,t)|\lambda\}$$

$$= \frac{1}{n!}(1 - e^{-\varepsilon t})^n e^{-t(\lambda - n\varepsilon)} \sum_{j=1}^{n} A(j,n)(\lambda/\varepsilon)^j. \tag{2.20}$$

The right hand side of (2.20) is now integrated against the gamma density (2.8) to give

$$\Pr\{X(t) = 0\} = \frac{\alpha^\nu}{(\alpha + t)^\nu}$$

$$\tag{2.21}$$

$$\Pr\{X(t) = n\} = \frac{1}{n!}(1 - e^{-\varepsilon t})^n e^{n\varepsilon t} \frac{\alpha^\nu}{\Gamma(\nu)} \sum_{j=1}^{n} \frac{A(j,n)}{\varepsilon^j} \frac{\Gamma(j+\nu,(\alpha+t)n\varepsilon)}{(\alpha+t)^{j+\nu}}$$

In this equation, $\Gamma(\mu,x)$ is the incomplete gamma function (14). It is defined by

$$\Gamma(\mu,x) = \int_x^\infty e^{-s} s^{\mu-1} \, ds \tag{2.22}$$

with power series and recursion relationship given by

$$\Gamma(\mu,x) = \Gamma(\mu) - x^\mu \Gamma(\mu) e^{-x} \sum_{n=0}^{\infty} \frac{x^n}{\Gamma(\mu+n+1)}$$

$$\tag{2.23}$$

$$\Gamma(\mu+1,x) = \mu\Gamma(\mu,x) + x^\mu e^{-x}$$

Equations (2.19), (2.21) and (2.23) completely specify the extended negative binomial distribution. Observe that for the binomial distribution given by (2.17), the coefficient of variation is

$$CV_B = \left( \frac{\varepsilon e^{-\varepsilon t}}{\lambda(1 - e^{-\varepsilon t})} \right)^{\frac{1}{2}} \tag{2.24}$$

The CV of the extended negative binomial distribution is, unfortunately, not easily calculated analytically. It can, however, be computed numerically quite easily. Figure 2.2 shows the ratio

$$\dot{R} = \frac{CV_{ENB}}{CV_B}$$

(for $t = 1$ and $\alpha = \lambda/\nu$) as a function of $\nu$. Figures 2.1 and 2.2 have virtually the same shape. At least qualitatively, then, the extended negative binomial distribution captures the idea of aggregation or patchiness in the same way that the negative binomial does when depletion is not important.

Returning to the binomial formula (2.17), observe that it has two interpretations. These are 1) the probability that $X(t) = n$, given $N = \lambda/\varepsilon$ and $p = 1 - e^{-\varepsilon t}$

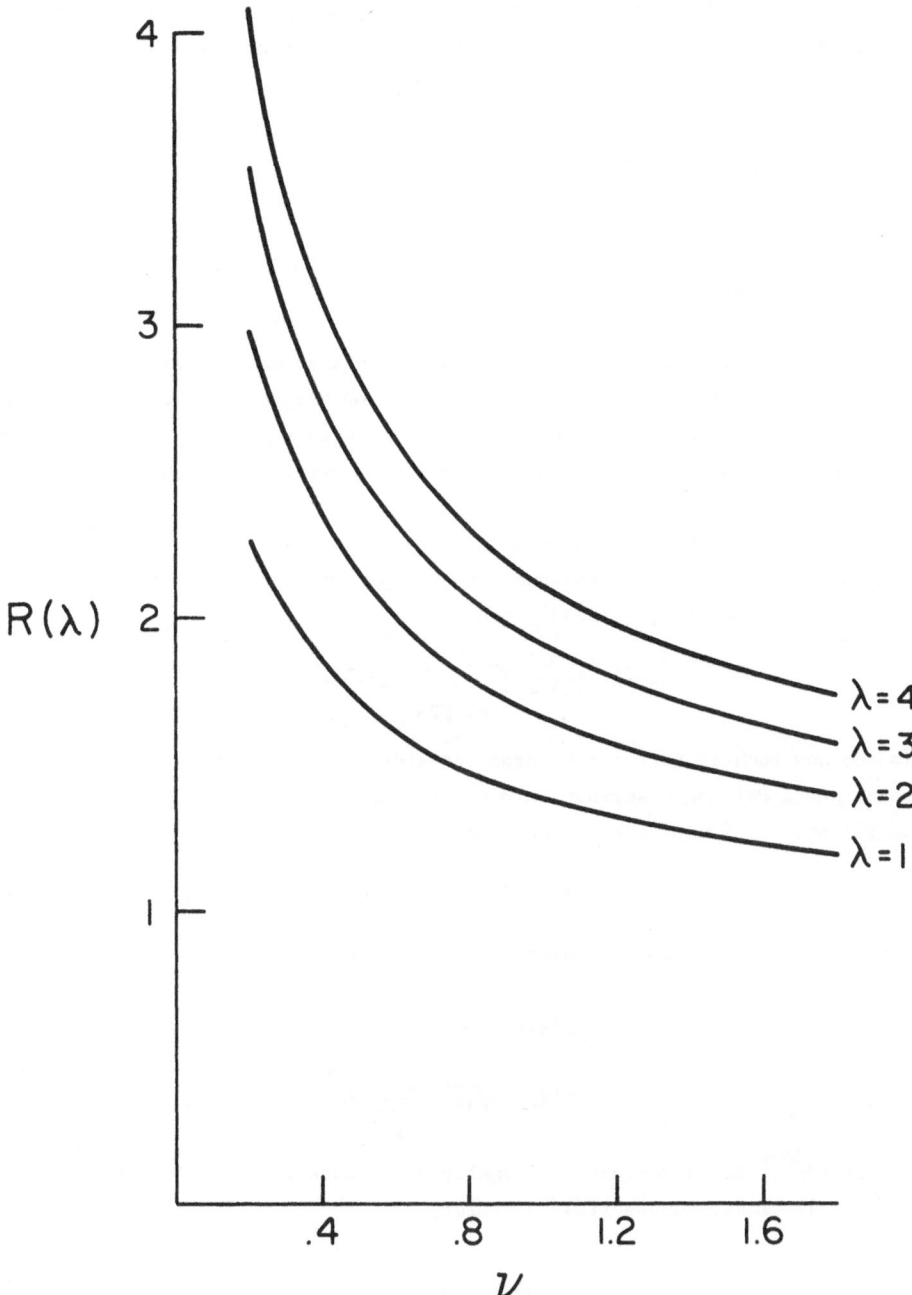

Fig. 2.2   Comparison of the Coefficients of Variation
of the Binomial and Extended Negative
Binomial Distributions.

and 2) the likelihood of N given n and p. By using the second interpretation
of (2.17), one can study the question of how many objects are present, given that
n were found (see, e.g. (15)). In the same way, (2.21) can be viewed as a likelihood for $\alpha$, given $\epsilon$, $\nu$, t and n (with the assumption that $\nu$ is "species characteristic"). Numerical investigations indicate that (2.21) is unimodal with a clearly
defined peak.

## 2.5. Transformations to Normality

The normal distribution is so ubiquitous and has so many inferential techniques
associated with it that it is worth asking if any of the previous four distributions
can be put into a form where the normal distribution is a good approximation. The
answer is yes, and the method proceeds as follows. First recall the central limit
theorem. Let $\{Z_i\}$ be a set of independent, identically distributed random variables
with mean $\mu_z$ and variance $\sigma_z^2$. Consider the quantity $U = [\sum_{i=1}^{n} Z_i - n\mu_z]/\sigma_z\sqrt{n}$.
According to the central limit theorem, U is approximately normally distributed
with mean 0 and variance 1, (N(0,1)), i.e.

$$\Pr\{U \leq u\} \approx \int_{-\infty}^{u} \frac{1}{\sqrt{2\pi}} e^{-s^2/2} \, ds. \tag{2.25}$$

To see how such an idea can be used, consider the simplest case of a variable
X(t) which has a Poisson distribution with parameter $\lambda t$. Consider a transformation
$Z = f(X(t))$ and a new variable Y defined by

$$Y = f(X(t)) - f(\lambda t) \tag{2.26}$$

A Taylor expansion gives (suppressing the t dependence of X)

$$Y \approx f'(\lambda t)[X - \lambda t]$$
$$= f'(\lambda t) \sqrt{\lambda t} \, \frac{[X - \lambda t]}{\sqrt{\lambda t}} \tag{2.27}$$

Since $[X-\lambda t]/\sqrt{\lambda t}$ is approximately normally distributed, Y will be too if
$f'(\lambda t) \sqrt{\lambda t} = 1$. Thus, choose f(x) to satisfy

$$f'(x) = \frac{1}{\sqrt{x}} \tag{2.28}$$

so that $f(x) = 2\sqrt{x}$. Thus, if X(t) is Poisson with parameter $\lambda t$, the transformed
variable $Y = 2\sqrt{X(t)} - 2\sqrt{\lambda t}$ is approximately N(0,1).

The same procedure can be followed if X(t) has a negative binomial distribution
with mean and variance given by

$$E\{X(t)\} = (\tfrac{\nu}{\alpha})t$$
$$\left.\text{Var}\{X(t)\} = (\tfrac{\nu}{\alpha})t + \tfrac{1}{\nu}(\tfrac{\nu}{\alpha}t)^2 \right\} \qquad (2.29)$$

Mimicking (2.26-2.28) shows that the appropriate $f(x)$ is given by

$$f(x) = \sqrt{\nu} \; \ln[2(\tfrac{x}{\nu} + \tfrac{x^2}{\nu^2}) + \tfrac{2x}{\nu} + 1] \qquad (2.30)$$

and then $Y = f(X(t)) - f((\nu/\alpha)t)$ is approximately $N(0,1)$.

For the binomial distribution, the calculation is a little trickier. Let X be binomially distributed with parameters N and p. Then proceeding as before gives

$$Y = f(X) - f(Np)$$

$$\approx f'(Np)[X-Np] \qquad (2.31)$$

$$= f'(Np)\sqrt{Np} \; \frac{[X-Np]}{\sqrt{Np(1-p)}} \; \sqrt{1-p} \; .$$

Setting $f(x) = 2\sqrt{x}$ gives

$$Y = \frac{[X-Np]}{\sqrt{Np(1-p)}} \; \sqrt{1-p}$$

so that Y is approximately $N(0,1-p)$. As an aside, it is worth noting that if one were interested in estimating p, i.e. in the quantity X/N, then the appropriate transformation to normality is $f(X/N) = \sin^{-1}(2(X/N)-1)$.

The advantage of such transformations to normality is that one can then apply existing theory for normally distributed random variables to the transformed variables. An example of such a use is Reed's work (16).


## 2.6. Diffusion Models

The last kinds of models introduced are ones in which the object being sought moves according to a diffusion process. That is, let $X(t) \in R^n$ (n=1,2,3) denote the position of an object moving randomly. Consider the increment $dX = X(t+dt) - X(t)$ and assume that given $X(t) = x$, dX is normally distributed with

$$E(dX) = b(x)dt + o(dt)$$
$$\left. E((dX)^2) = a(x)dt + o(dt) \right\} \qquad (2.32)$$

The functions $b(x)$ and $a(x)$ are called the infinitesimal mean and variance respectively (see, e.g. (13)). Consider the density $f(x,t)$ defined by

$$f(x,t)dx = \Pr\{X(t) \in (x,x+dx)\}. \tag{2.33}$$

It is assumed that $f(x,0) = f_0(x)$ is given. It can be shown (e.g. (12) or (13))
that $f(x,t)$ satisfies the "forward" partial differential equation

$$\frac{\partial f}{\partial t} = \frac{1}{2} \sum_{i,j} \frac{\partial^2}{\partial x_i \partial x_j} (a_{ij}(x)f) - \sum_i \frac{\partial}{\partial x_i} (b_i(x)f) \tag{2.34}$$

where $(a_{ij}) = a$ and $(b_1,b_2) = b$.

It is also instructive to consider a "backward" equation, which arises as follows. Let A be a specified region and $u(x,t,T)$ be defined by

$$u(x,t,T) = \Pr\{X(T) \in A | X(t) = x\}. \tag{2.35}$$

The equation that $u(x,t,T)$ satisfies can be derived in a straightforward manner by
noting that

$$u(x,t,T) = \int u(x+z,t+dt,T)\Pr\{dX = z | X(t) = x\}dz$$

$$= \int \left[ u(x,t,T) + \frac{\partial u}{\partial t} dt + \sum_i z_i \frac{\partial u}{\partial x_i} + O(z^3) \right. \tag{2.36}$$

$$\left. + \frac{1}{2} \sum_{i,j} z_i z_j \frac{\partial^2 u}{\partial x_i \partial x_j} \right] \Pr\{dX = z | X(t) = x\}dz.$$

Taking expectations in (2.36) and using (2.32) gives

$$u(x,t,T) = u(x,t,T) + \frac{\partial u}{\partial t} dt + \sum_i b_i(x) \frac{\partial u}{\partial x_i} dt$$

$$+ \frac{1}{2} \sum_{i,j} a_{ij}(x) \frac{\partial^2 u}{\partial x_i \partial x_j} dt + o(dt), \tag{2.37}$$

so that $u(x,t,T)$ satisfies

$$0 = u_t + \sum_i b_i(x) \frac{\partial u}{\partial x_i} + \frac{1}{2} \sum_{i,j} a_{ij}(x) \frac{\partial^2 u}{\partial x_i \partial x_j} . \tag{2.38}$$

The initial condition for (2.34) is $f(x,0) = f_0(x)$. The final condition for
(2.38) is

$$\lim_{t \to T} u(x,t,T) = \begin{cases} 1 & \text{if } x \in A \\ 0 & \text{otherwise.} \end{cases}$$

Depending upon the particular problem, boundary conditions may also be needed.

In the next section, the effect of search on the time evolution of $f(x,t)$ or $u(x,t,T)$ will be discussed.

## 3. Random Search Models

This section contains a description of a class of models that date back to WW II. These models have proved extremely useful in a number of contexts (see, e.g. (17), (18), (19)) and a careful study of them is warranted. The phrase "random search" is actually inappropriate, but has stuck with the class of models for 40 years.

## 3.1. Random Search Formula

To begin, imagine once again a region $Q$ of area $A$ that contains exactly one object. Let $P(t)$ denote the probability of detecting the object after $t$ hours of search. Set $Q(t) = 1 - P(t)$. The first way to derive the random search formula is by assuming a lack of learning or memorylessness on the part of the searcher. That is, assume that

$$Q(s+t) = Q(s)Q(t). \qquad (3.1)$$

If equation (3.1) is valid, then $Q(t)$ must have the form $Q(t) = e^{-\beta t}$ (see, e.g. (20)) so that

$$P(t) = 1 - e^{-\beta t}. \qquad (3.2)$$

This is called the random search formula. There are two difficulties with formula (3.2). The first is that one must assume a lack of learning, embodied in the assumption (3.1). The second is that $\beta$ is unknown. Koopman's (17) original method for deriving the random search formula provides a direct way to get at $\beta$. To use it, introduce a sweep or detection width $W$ which has the definition (17) that

$$\text{Pr\{not detecting an object within } W\}$$
$$= \text{Pr\{detecting an object outside of } W\}. \qquad (3.3)$$

Now imagine that the search speed is $v$ so that the total search track has length $L = vt$. Divide this track into $n$ segments of length $vt/n$. Assume that these tracks are placed down randomly over $Q$ (hence the name random search). The probability of detection on any single track is assumed to be $(\frac{Wvt}{n})/A$, i.e., that the object sought is uniformly distributed in $A$. Finally, assuming that detections are independent random variables gives

$$P(t) = 1 - (1 - \frac{Wvt}{n} \cdot \frac{1}{A})^n. \tag{3.4}$$

Letting $n \to \infty$ in (3.4) gives

$$P(t) = 1 - e^{-Wvt/A}. \tag{3.5}$$

Before discussing the implications of (3.5), one more derivation will be presented. To do this, observe that the assumptions leading to (3.5) are the same as

$$Pr\{\text{detection in } (t,t+dt)\} = \frac{Wv\,dt}{A} + o(dt). \tag{3.6}$$

Thus

$$\begin{aligned} Q(t+dt) &= Pr\{\text{no detection up to } t + dt\} \\ &= Q(t)(1 - \frac{Wv\,dt}{A}) + o(dt). \end{aligned} \tag{3.7}$$

Equation (3.7) leads to the differential equation

$$\frac{dQ}{dt} = -\frac{Wv}{A} Q(t). \tag{3.8}$$

The initial condition $Q(0) = 1$ gives $Q(t) = e^{-Wvt/A}$, which is the same as (3.5).

According to (3.5), the time to detection is exponentially distributed with parameter $\lambda = Wv/A$ and the mean time to detection is given by $A/Wv$. If there are $k$ searchers rather than one, and they search independently, then (3.6) shows that $W$ in (3.5) should be replaced by $kW$. If there are $N$ objects which are "removed" (e.g. schools of fish are harvested, plants with pests are tagged) then

$$Pr\{\text{detecting } k \text{ of } N \text{ objects}\} = \binom{N}{k}(1 - e^{-Wvt/A})^k (e^{-Wvt/A})^{N-k}. \tag{3.9}$$

If $N$ is large and $t$ is small, (3.9) can be replaced by the Poisson approximation to the binomial. In that case, the Poisson parameter is $N(1 - e^{-Wvt/A})$ which is approximately $NWvt/A$ if $t$ is small.

The basic random search formula (3.2) seems to hold in a variety of situations in which one would expect it not to be valid. Some striking examples are discussed by Washburn (19) in which two humans play a computer game of searcher and hider, and yet (3.2) seems to be valid (ref. (19), pg. 2.9). If the object being sought moves, then very often an "exhaustive" search can be turned into a "random" search by random or quasi-random motion of the object being sought.

## 3.2. Diffusion Models With Search (21,22,23)

One extension of these ideas is to consider a randomly moving object as in the

previous section and ask for the effect of search on its probability density. To do
this, rather than using sweep width, introduce a detection function $\psi(x,t,z)$ defined
by

$$\psi(x,t,z)dt = \Pr\{\text{detection in } (t,t+dt)|X(t) = x,$$

$$\text{searcher is at } z\}. \tag{3.10}$$

One is now interested in the joint density, $f(x,t,z)$, for location of the object
sought and unsuccessful search along the path $Z(\tau)$, $0 \leq \tau \leq t$. It can be shown that
$f(x,t,z)$ satisfies (21, 23)

$$\frac{\partial f}{\partial t} = \frac{1}{2} \sum_{i,j} \frac{\partial^2}{\partial x_i \partial x_j} (a_{ij}f) - \sum_i \frac{\partial}{\partial x_i} (b_i f) - \psi(x,t,z)f. \tag{3.11}$$

Observe that the probability of detection by time $T$, $P_D(T)$, is given by

$$P_D(T) = 1 - \int f(x,T,Z)dx, \tag{3.12}$$

since the integral on the right hand side is the probability of unsuccessful search.

Another way of getting to the probability of detection is to consider it directly. That is, let

$$u(x,t,T) = \Pr\{\text{no detection in } (t,T)|X(t) = x\}. \tag{3.13}$$

Then the law of total probabilities gives

$$u(x,t,T) = (1 - \psi(x,t,Z)dt)E_{dX}(u(x+dX,t+dt,T)). \tag{3.14}$$

A Taylor expansion of (3.14) gives

$$u(x,t,T) = (1-\psi(x,t,Z)dt)E_{dX}\Big( u(x,t,T) + \frac{\partial u}{\partial t} dt$$

$$+ \sum_i dX_i \frac{\partial u}{\partial x_i} + \frac{1}{2} \sum_{i,j} dX_i dX_j \frac{\partial^2 u}{\partial x_i \partial x_j} + o(\|dX\|^3)\Big). \tag{3.15}$$

Using the normality assumption on $dX$ and the moments (2.32) leads to the equation

$$0 = \frac{\partial u}{\partial t} + \sum_i b_i \frac{\partial u}{\partial x_i} + \frac{1}{2} \sum_{i,j} a_{ij} \frac{\partial^2 u}{\partial x_i \partial x_j} - \psi(x,t,Z)u. \tag{3.16}$$

This equation satisfies the end condition $u(x,T,T) = 1$. (It is worth noting that
(3.16) is the differential equation satisfied by the Feynman-Kac functional (21,24)

$$u(x,t,T) = E_x\left[\exp\left\{-\int_t^T \psi(X(s),s,Z)dS\right\}\Big|X(t) = x\right] \qquad (3.17)$$

where $E_x$ denotes the expectation over all paths of $X(t)$.)

Although (3.11) and (3.16) may appear to be arcane, they are actually quite useful as the example in Section 6 shows.

### 3.3. Random Search With Harvest Time

Return now to the simple random search formula, valid when N schools of fish are present. That formula shows that the time to detecting a school is exponentially distributed with parameter NWv/A. Suppose now that each detection leads to a period of no search. This would occur, for example, if after a detection occurs there is a harvest time of $\tau$. How many schools can one expect to find in a given period T and what kind of distribution does the catch have?

Common sense suggests the following. Let M(T) denote the number of schools completely or partially harvested in T and let $\overline{M}(T)$ be its expectation. Then one would anticipate

$$\overline{M}(T) \approx \frac{\text{Total time}}{\text{Average time per encounter}} = \frac{T}{(A/NWv) + \tau} . \qquad (3.18)$$

Further study is required, however, to make the sense in which the approximation holds clearer.

The process described "search-harvest-search-harvest" is a renewal process ((12),(13)) since after each harvest the operation begins again, with exactly the same kind of distribution. There is a considerable literature associated with renewal theory, but a brief introduction to it here is worthwhile (also see (2), where the fishing problem is treated with false detections). Let F(t) be defined by

$$F(t) = \Pr\{\text{first renewal event (i.e.}$$
$$\text{"search-harvest") is completed before t}\} \qquad (3.19)$$

and let $f(t) = dF/dt$ be the probability density for the first renewal event. Actually, F(t) and f(t) hold for any renewal event, because of the assumption about identical distributions. One can then derive an equation for $\overline{M}(T)$. In particular assume that the first event is completed at y. Then

$$\overline{M}(T) = E_y\{1 + \overline{M}(T-y)\} \qquad (3.20)$$

where $E_y$ is taken using (3.19). That is

$$\bar{M}(T) = \int_0^T \bar{M}(T-y)f(y)dy + F(T). \tag{3.21}$$

This is the famous "renewal equation". Assuming that $T \gg \tau$, $A/NWv$ one can seek a solution of (3.21) in the form

$$\bar{M}(T) = K_1 T + K_2 + K_3/T + \cdots \tag{3.22}$$

where the $K_i$ are constants. That is, one substitutes (3.22) into (3.21) and sets coefficients of various powers of $T$ equal to zero. When this is done, one finds, with $NWv/A = \lambda$

$$K_1 = \frac{1}{\frac{1}{\lambda} + \tau} \tag{3.23}$$

$$K_2 = -\left\{ \frac{(\lambda\tau+1)^2 - 1}{2(\lambda\tau+1)^2} \right\}. \tag{3.24}$$

Equations (3.22-3.24) show how the correction to (3.18) is obtained.

What about the distribution of M(T)? The precise distribution of M(T) is actually quite difficult to calculate, but some approximations are useful. In particular, if $T$ is large, then one can appeal to the central limit theorem for renewal processes (see, e.g. (12)). According to this theorem, if $\mu$ and $\sigma^2$ are the mean and variance of the first renewal event, then

$$Z = \frac{M(T) - T/\mu}{(T\sigma^2/\mu^3)^{\frac{1}{2}}} \tag{3.25}$$

is approximately normally distributed with mean zero and variance 1. For the problem at hand, one easily finds that

$$\left. \begin{aligned} \mu &= \frac{1}{\lambda} + \tau \\ \sigma^2 &= \frac{1}{\lambda^2} \end{aligned} \right\} \tag{3.26}$$

so that for this problem the approximate asymptotic distribution of the catch is known.

## 4. Exhaustive Search Models

The opposite extreme of random search is exhaustive search, according to which all objects within the sweep width 2W of the searcher are detected with probability one. In this kind of problem, one has what is known as a "cookie-cutter" detection function, since

$$\psi(x,t,z)dt = \begin{cases} 1 & \text{if the distance between} \\ & x \text{ and } z \leq W \\ \\ 0 & \text{otherwise} \end{cases} \tag{4.1}$$

The problem of exhaustive search in fisheries was studied many years ago by J. Neyman (10). In addition to being published in a sometimes hard to find publication, Neyman's paper is hard to read (at least in this author's opinion). In this section, a new derivation of Neyman's work is presented, followed by a number of new extensions.

To start, imagine once more a region $\Omega$ of area $A$ that contains a total of $M$ schools of fish. A fishing vessel operates in this region, detects everything within $2W$ of the vessel path, and spends $h$ hours of harvest per detection. Imagine now that a time interval $[0,T]$ is given. One can ask two general questions:

1) What is the catch in $[0,T]$?

2) What is the vessel doing (fishing or searching) at time $T$?

These are simple questions, but they require complicated answers.

Let $A_k$ denote the area that is searched in $T - kh$ hours and let $n$ be the integer part of $T/h$, written as $n = [T/h]$. Neyman proved three propositions about the two questions that were posed above. In a different order (for pedagogical reasons) these are the following.

1) The probability that the catch is $n = [T/h]$ and the vessel is fishing at time $T$ is equal to the probability that $A_n$ contains at least $n+1$ schools. To see this, simply observe that if $A_n$ contains at least $n+1$ schools then $n$ schools are captured in elapsed time $nh$, and another school is detected but not completely harvested in the remaining time $T - nh$. Thus

$$\Pr\{\text{catch is } n = [T/h], \text{ fishing at } T\}$$

$$= \Pr\{A_n \text{ contains at least } n+1 \text{ schools}\} \tag{4.2}$$

$$= 1 - \sum_{m=0}^{n} \Pr\{A_n \text{ contains } m \text{ schools}\}.$$

2) For $k$ such that $kh < T$, the probability that the catch in $[0,T]$ is $k$ schools and the vessel is searching at time $T$ is found as follows. The vessel spends $kh$ hours fishing and in the remaining time does not find another school. The total area swept in this problem is thus $A_k$, which must contain exactly $k$ schools. Thus

$$\text{Pr\{catch is } k \text{ schools, searching at } T\}$$

$$= \text{Pr\{}A_k \text{ contains exactly } k \text{ schools\}.} \tag{4.3}$$

3)    The probability that the catch is $k$ schools in $[0,T]$ and the vessel is fishing at $T$ is computed as follows.  For this event to occur, the region $A_k$ must contain at least $k+1$ schools <u>and</u> the (smaller) region $A_{k+1}$ can contain no more than $k$ schools.  That is, up to $k$ schools are found in the region $A_{k+1}$ and then the vessel starts searching.  The extra schools are found in the little bit of area $A_k - A_{k+1}$.  The following picture will be helpful in the calculations that must be done.

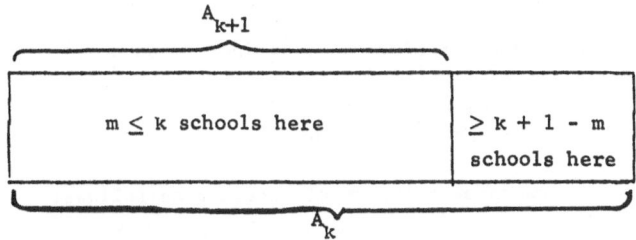

Thus

$$\text{Pr\{catch } k \text{ schools completely in } (0,T), \text{ catching the}$$
$$k+1^{st} \text{ at } T\}$$

$$= \text{Pr\{}A_k \text{ contains at least } k+1 \text{ schools, } A_{k+1} \text{ contains}$$
$$\quad \text{at most } k \text{ schools\}}$$

$$= \sum_{m=0}^{k} \text{Pr\{}A_{k+1} \text{ contains } m \text{ schools\}Pr\{}A_k - A_{k+1} \text{ contains}$$
$$\quad \text{at least } k+1-m \text{ schools} | m \text{ schools in } A_{k+1}\}$$

$$= \sum_{m=0}^{k} \text{Pr\{}A_{k+1} \text{ contains } m \text{ schools\}[1 - Pr\{}A_k - A_{k+1}$$
$$\quad \text{contains} \le k-m \text{ schools} | A_{k+1} \text{ contains } m \text{ schools\}} ]$$

$$= \sum_{m=0}^{k} \text{Pr\{}A_{k+1} \text{ contains } m \text{ schools\}}$$
$$\quad - \sum_{m=0}^{k} \text{Pr\{}A_k \text{ contains } m \text{ schools\}.} \tag{4.4}$$

Now one can particularize the distributions of interest.  The fundamental distribution is the binomial:

$$\text{Pr}\{\text{region of size } W \text{ has } m \text{ schools}\}$$

$$= \binom{M}{m} (\frac{W}{A})^{m} (1 - \frac{W}{A})^{M-m} . \tag{4.5}$$

When M, A $\to \infty$ in such a way that M/A = $\lambda$ is constant (4.5) has the limiting form

$$\text{Pr}\{\text{region of size } W \text{ contains } m \text{ schools}\}$$

$$= \frac{e^{-\lambda W} (\lambda W)^{m}}{m!} . \tag{4.6}$$

The natural generalization of (4.6) is to assume that $\lambda$ has a gamma distribution with parameters $(\nu, \alpha)$, leading to a negative binomial distribution:

$$\text{Pr}\{\text{region of size } W \text{ contains } m \text{ schools}\}$$

$$= \binom{m+\nu-1}{m} (\frac{\alpha}{\alpha+W})^{\nu} (\frac{W}{\alpha+W})^{m} . \tag{4.7}$$

Next, define the two probabilities of interest as follows:

$$P_f(k,T) = \text{Pr}\{\text{catch is } k \text{ schools and the vessel}$$
$$\text{is fishing at } T\}$$
$$\tag{4.8}$$
$$P_s(k,T) = \text{Pr}\{\text{catch is } k \text{ schools and the vessel}$$
$$\text{is searching at } T\}$$

For the binomial distribution, these quantities become

$$P_f(n,T) = 1 - \sum_{m=0}^{n} \binom{M}{m} (\frac{2W(T-nh)}{A})^{m} (1 - \frac{2W(T-nh)}{A})^{M-m} . \tag{4.9a}$$

$$P_f(k,T) = \sum_{m=0}^{k} \binom{M}{m} (\frac{2W(T-(k+1)h)}{A})^{m} (1 - \frac{2W(T-(k+1)h)}{A})^{M-m}$$

$$- \sum_{m=0}^{k} \binom{M}{m} (\frac{2W(T-kh)}{A})^{m} (1 - \frac{2W(T-kh)}{A})^{M-m} \tag{4.9b}$$

$$k = 0,1,2,\ldots,(n-1).$$

$$P_s(k,T) = \binom{M}{k} (\frac{2W(T-kh)}{A})^{k} (1 - \frac{2W(T-kh)}{A})^{M-k} \tag{4.9c}$$

$$k = 0,1,2,\cdots$$

For the Poisson distribution, these quantities become

$$P_f(n,T) = 1 - \sum_{m=0}^{n} e^{-\lambda(2W(T-nh))} \frac{(\lambda 2W(T-nh))^m}{m!}. \qquad (4.10a)$$

$$P_f(k,T) = \sum_{m=0}^{k} e^{-\lambda(2W(T-(k+1)h))} \frac{(\lambda 2W(T-(k+1)h))^m}{m!}$$

$$- \sum_{m=0}^{k} e^{-\lambda(2W(T-kh))} \frac{(\lambda 2W(T-kh))^m}{m!} \qquad (4.10b)$$

$$k = 0,1,2,\ldots,n-1.$$

$$P_s(k,T) = e^{-\lambda(2W(T-kh))} \frac{(\lambda 2W(T-kh))^k}{k!}$$

$$\qquad (4.10c)$$

$$k = 0,1,2,\cdots$$

Finally, for the negative binomial, these quantities become

$$P_f(n,T) = 1 - \sum_{m=0}^{n} \binom{m+\nu-1}{m} \left(\frac{\alpha}{\alpha+2W(T-nh)}\right)^\nu \left(\frac{2W(T-nh)}{\alpha+2W(T-nh)}\right)^m. \qquad (4.11a)$$

$$P_f(k,T) = \sum_{m=0}^{k} \binom{m+\nu-1}{m} \left(\frac{\alpha}{\alpha+2W(T-(k+1)h)}\right)^\nu \left(\frac{2W(T-(k+1)h)}{\alpha+2W(T-(k+1)h)}\right)^m$$

$$- \sum_{m=0}^{k} \binom{m+\nu-1}{m} \left(\frac{\alpha}{\alpha+2W(T-kh)}\right)^\nu \left(\frac{2W(T-kh)}{\alpha+2W(T-kh)}\right)^m \qquad (4.11b)$$

$$k = 0,1,2,\ldots,n-1.$$

$$P_s(k,T) = \binom{k+\nu-1}{k} \left(\frac{\alpha}{\alpha+2W(T-kh)}\right)^\nu \left(\frac{2W(T-kh)}{\alpha+2W(T-kh)}\right)^k. \qquad (4.11c)$$

In (4.11), the parameters $\nu$ and $\alpha$ are related to N as follows. Observe that (4.11) can be obtained by integrating (4.10) against a gamma density with parameters $\nu$ and $\alpha$. Thus, one would assume $E\{\lambda\} = \nu/\alpha$. But $\lambda = M/A$, so that one can leave $\nu$ as an adjustable parameter and pick $\alpha$ by setting $\nu/\alpha = M/A$ or $\alpha = \nu A/M$.

These formulas are all easily evaluated on a desk-top microcomputer. From these, one can, of course, make theoretical projections about future catches.

Perhaps a more important question, however, is the use of data to estimate the actual number of schools  M  in the region or the density $\lambda$ = M/A of schools.  To do this, assume that one knows

$$Z(T) = \text{Number of schools completely or} \qquad (4.12)$$
$$\text{partially fished by } T.$$

The distribution of $Z(T)$ is easily computed:

$$\Pr\{Z(T) = k\} = P_s(k,T) + P_f(k-1,T) \equiv P_Z(k,T). \qquad (4.13)$$

Now for a particular model, one can think of $P_Z(k,T)$ as the likelihood of some of the parameters, given  k  and  T.  In particular, assume that the sweep width  W  and the area of the region  A  are determined independently of the data set being analyzed.  For example, the Coast Guard publishes tables of sweep width as a function of environmental conditions and target size (25).  The size of the region being searched can often be estimated independent of the data set as well (15).  With this approach, one has the set  (k,T,W,A)  and wants to estimate the unknown parameters in (4.9-4.11).  In particular

| Model | Parameter to be Estimated |
|---|---|
| Binomial | M |
| Poisson | $\lambda$ |
| Negative Binomial | $\alpha$ |

For the negative binomial case, one must assume that the value of  $\nu$  is known. This could be done, for instance, if there is historical data.  Then one can measure $\nu$  either as  i) the parameter relating the mean and variance in the overall historical catch rate (Var = mean + $\frac{1}{\nu}$(mean)$^2$) or  ii) the coefficient of variation of the local catch rate at one spot.  In either case, the value of  $\nu$  can be estimated. Thus, for a particular model, given the catch  k  in operating time  T,  one can estimate the number of schools, the density, or the negative binomial parameter  $\alpha$.

Once maximum likelihood estimates of the parameter are found, one can construct confidence regions by summing the probabilities in the appropriate model.  An example of this approach, for random search, is found in (15).

## 5.  Catch Per Unit Effort and Stock Abundance

This and the next section contain two examples of how the methods of Sections 1 - 4 can be applied.  Further details for each of the problems described can be found in (26) or (27).

A large number of marine fisheries are managed under the hypothesis that catch per unit effort (CPUE) is proportional to abundance (N)

$$CPUE = q \cdot N \qquad (5.1)$$

where the constant $q$ is called the catchability coefficient. There is growing empirical evidence, however, (28,29) that the relationship may be

$$CPUE = AN^p \qquad (5.2)$$

where $0 < p < 1$. If this is the case, there may be serious consequences. For example, if a time series of CPUE shows a decline, then the population has declined even more dramatically, if (5.2) is valid. If the time series of CPUE oscillates, then the true population oscillates even more. Finally, equating (5.1) and (5.2) shows that

$$q = AN^{p-1} \qquad (5.3)$$

which increases as $N$ decreases. This corresponds to "increased catchability at low population levels". The purpose of this section is to show how (5.2) may arise when searching and handling are taken into account. There are two approaches to this problem. The first, which is somewhat simpler, assumes that catch is fixed and operating time is random. The second assumes that operating time is fixed and catch is random.

The simplest theory is an elaboration of work by Paloheimo and Dickie (30), Beddington (31), and Cooke (32). It goes as follows. Imagine a total operating time $T$ composed of search time $T_s$ and handling time $T - T_s$. Assume that catch is a Poisson process with parameter $\lambda = qN$ (this model thus ignores depletion of the stock. Similar results are obtained, however, when one takes depletion into account, see (26)). Specify a catch $C$ and assume that each detection results in a handling time of $h$ hours. Now the time for the first $C$ events of a Poisson process with parameter $\lambda$ follows a gamma distribution with parameters $C$ and $\lambda$. The operating time $T$ needed to achieve the catch $C$ is thus a random variable with first two moments

$$E\{T\} = E\{T_s\} + Ch = C/\lambda + Ch$$
$$= \frac{C}{qN} + Ch = C\left(\frac{1+qhN}{qN}\right) \qquad (5.4)$$

$$Var\{T\} = Var\{T_s\} = \frac{C}{\lambda^2} = \frac{C}{(qN)^2} . \qquad (5.5)$$

Consider the catch per unit effort CPUE defined by

$$CPUE = \frac{C}{T} . \qquad (5.6)$$

Paloheimo and Dickie, Beddington, and Cooke all work with the expectation of CPUE. If $\bar{T} = E\{T\}$, it is approximately found as follows:

$$E\{C/T\} = CE\{1/T\}$$

$$= CE\left\{\frac{1}{\bar{T}} - \frac{1}{\bar{T}^2}(T-\bar{T}) + \frac{2}{\bar{T}^3}(T-\bar{T})^2 + \cdots\right\} \qquad (5.7)$$

$$= C\left\{\frac{1}{\bar{T}} + \frac{2}{\bar{T}^3}\,\text{Var}\{T\} + \cdots\right\}$$

Substituting from (5.4) and (5.5) gives

$$E\{CPUE\} = \frac{qN}{1+qhN} + \frac{2}{C(qN)^2}\left\{\frac{qN}{1+qhN}\right\}^3 + \cdots \qquad (5.8)$$

Assuming that the first term dominates all others, one has

$$E\{CPUE\} \approx \frac{qN}{1+qhN} \, , \qquad (5.9)$$

which saturates with $N$, but certainly does not exhibit $AN^p$ behavior.

Cooke (32) proposes that $q$ should actually have a distribution itself, since it depends upon a multitude of operational, environmental and human factors. He suggests that the distribution should be skewed and uses a log-normal distribution on $q$; another choice is the gamma density with parameters $\nu$ and $\alpha$. Averaging (5.9) over this density gives

$$\langle E\{CPUE\}\rangle_q = \frac{N\alpha^\nu}{\Gamma(\nu)} \int_0^\infty \frac{q^\nu e^{-\alpha q}}{1+qhN}\,dq \qquad (5.10)$$

$$= \frac{\nu}{h}\left(\frac{\alpha}{hN}\right)^\nu \Gamma(-\nu, \alpha/hN)e^{\alpha/hN} \qquad (5.11)$$

where $\Gamma(s,x)$ is the incomplete gamma function, defined by (14)

$$\Gamma(s,x) = \int_x^\infty e^{-t}t^{s-1}\,dt. \qquad (5.12)$$

It has the following limiting behavior:

$$\lim_{x \to 0} \Gamma(s,x) = \Gamma(s)$$

$$\left.\vphantom{\begin{array}{c}1\\1\\1\end{array}}\right\} \qquad (5.12)$$

$$\Gamma(s,x) \sim x^{s-1}e^{-x} \qquad \text{large } x$$

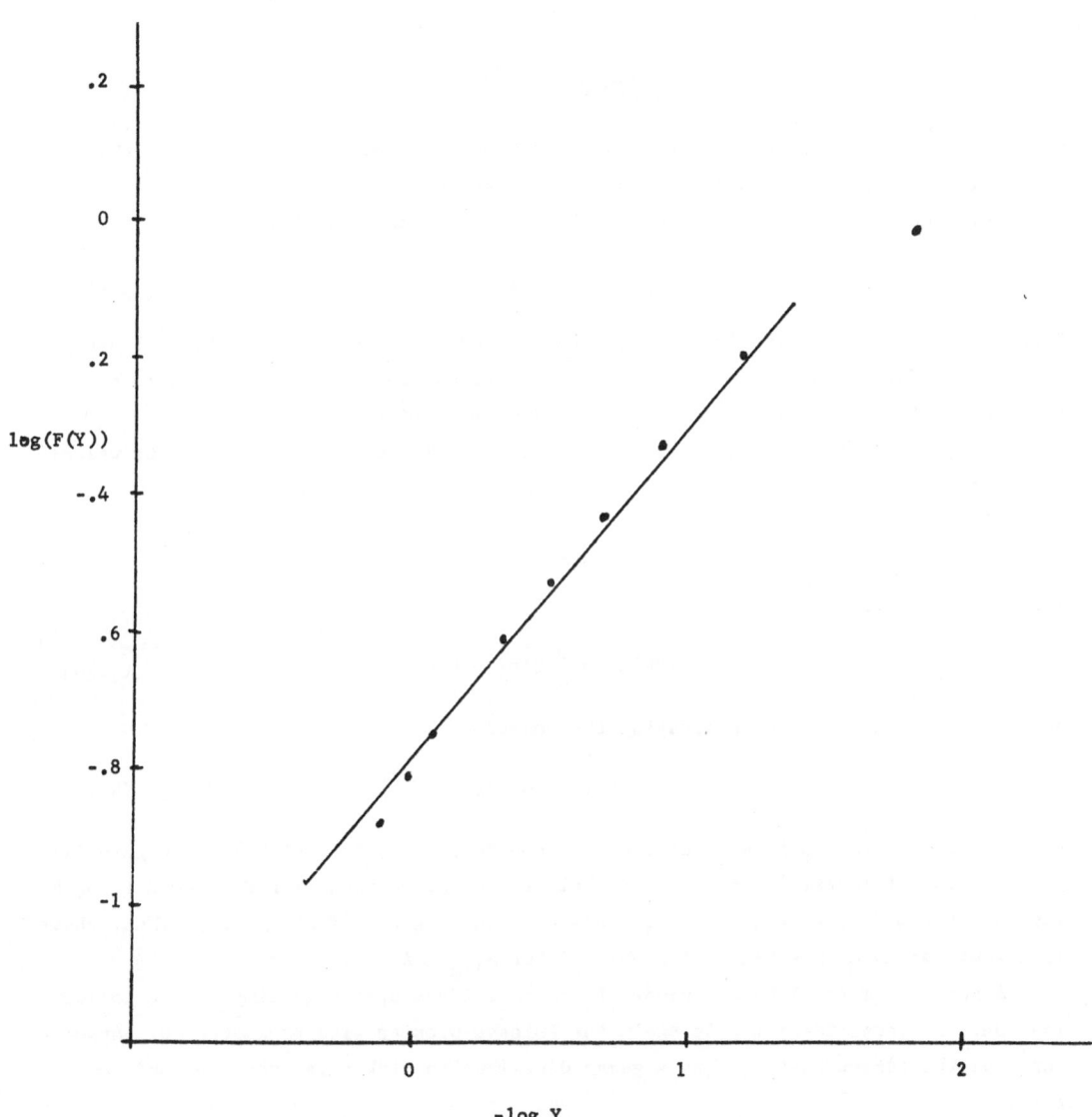

Fig. 5.1    A Plot of log(CPUE) vs log N
            in the First Model.

Use of the second limit behavior in (5.11) shows that as $\alpha/hN \to \infty$, which corresponds to $h \to 0$, one finds

$$\langle E\{CPUE\}\rangle_q \to \frac{N\nu}{\alpha} \, . \tag{5.13}$$

That is, without handling time, the averaged CPUE is proportional to $N$ and the proportionality constant is simply the average value of $q$.

For $\alpha/hN \to 0$, which corresponds to $N \to \infty$, one finds from (5.10) that

$$\langle E\{CPUE\}\rangle_q \to 1/h \, . \tag{5.14}$$

That is, as the population level increases the CPUE approaches the reciprocal handling time. This is due to search being insignificant for large population levels -- schools are found immediately and all the time is spent harvesting.

Thus the limiting forms of (5.11) make sense. To study the intermediate values, define a dimensionless parameter $Y$ by

$$Y = \frac{\alpha}{hN} \, . \tag{5.15}$$

Then one can write

$$\langle E\{CPUE\}\rangle_q = \frac{\nu}{h} Y^\nu \Gamma(-\nu, Y) e^Y \tag{5.16}$$

so that one is interested in studying the behavior of

$$F(Y) = Y^\nu \Gamma(-\nu, Y) e^Y \tag{5.17}$$

as $Y$ varies. To within a multiplicative constant, $F(Y)$ is $\langle E\{CPUE\}\rangle_q$. Figure 5.1 shows a plot of $\log(F(Y))$ vs $-\log Y$ (recall $Y = \alpha/hN$, so that $-\log Y = \log N + \log h - \log \alpha$). For moderate values of $Y$, this plot has a slope of about 0.5. Thus, there is a range of abundance values for which $\langle E\{CPUE\}\rangle_q \approx AN^{.5}$.

A somewhat more elaborate theory based on a fixed operating time is the following one. The starting point is again the Poisson process with parameter $qN$. Assuming from the outset that $q$ has a gamma distribution with parameters $\nu$ and $\alpha$ gives

$$\Pr\{n \text{ encounters in search time } t\} = \int_0^\infty \frac{e^{-qNt}}{n!} (qNt)^n \frac{e^{-\alpha q} \alpha^\nu q^{\nu-1}}{\Gamma(\nu)} \, dq \tag{5.18}$$

$$= \binom{n+\nu-1}{n} \left(\frac{\alpha}{\alpha+Nt}\right)^\nu \left(\frac{Nt}{\alpha+Nt}\right)^n,$$

the familiar negative binomial distribution. Now, the possible events are catches of $0, 1, 2, \cdots N^* = \min(N, [T/h])$ schools, where $T$ is the total operating time. The

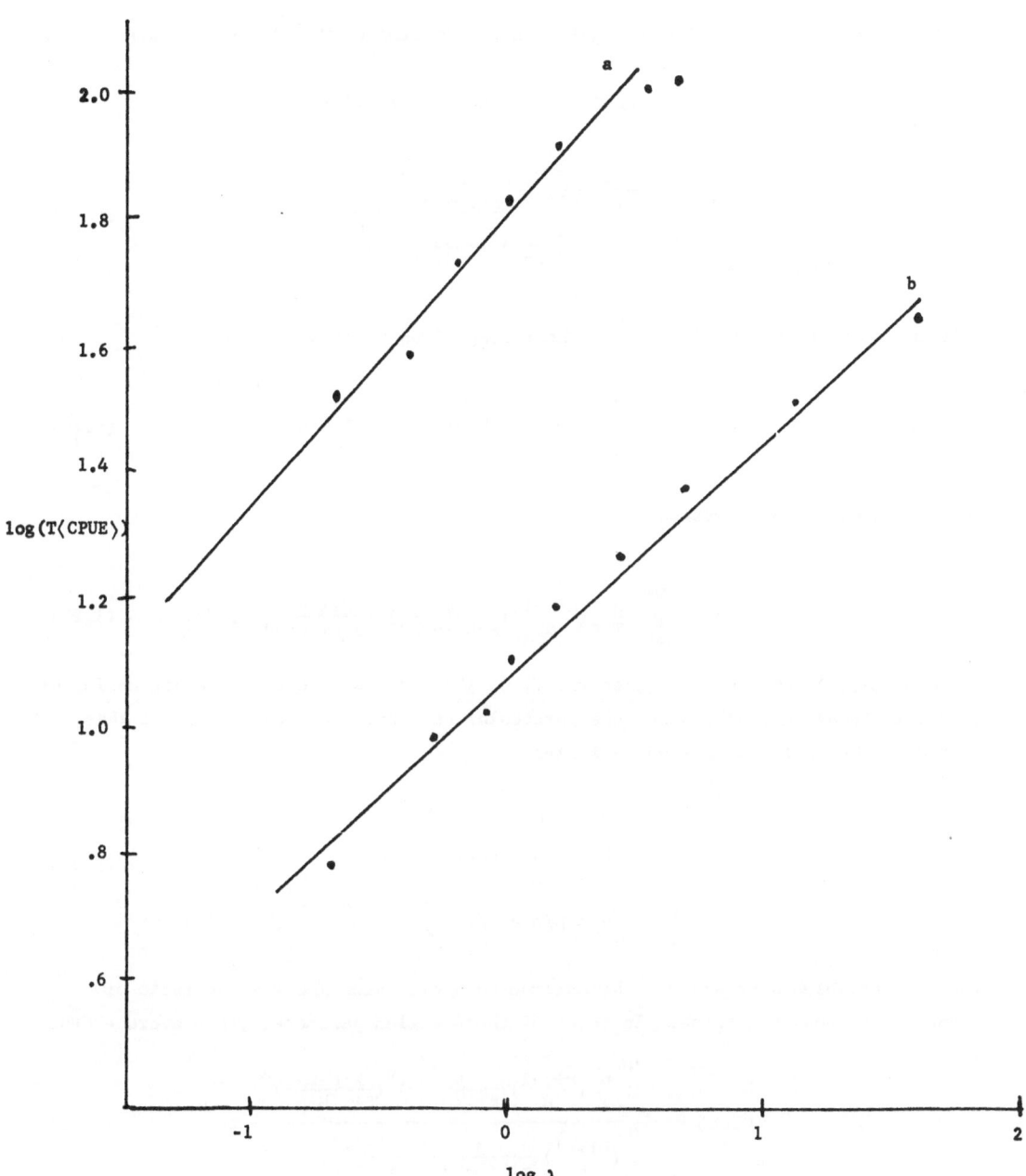

Fig. 5.2    Plot of log(T⟨CPUE⟩) vs log λ.
a: M = 20,   b: M = 10.

event of a catch of  n  schools requires a search time of T - nh hours.  Thus

Pr{catch of n schools in operating time T} =

$$\frac{\binom{n+\nu-1}{n}\left(\frac{\alpha}{\alpha+N(T-nh)}\right)^{\nu}\left(\frac{N(T-nh)}{\alpha+N(T-nh)}\right)^{n}}{\sum\limits_{k=0}^{N*}\binom{k+\nu-1}{k}\left(\frac{\alpha}{\alpha+N(T-kh)}\right)^{\nu}\left(\frac{N(T-kh)}{\alpha+N(T-kh)}\right)^{k}} \tag{5.19}$$

Call the denominator in (5.19)  $\mathfrak{D}$.  The average CPUE is then

$$\langle CPUE \rangle = \sum_{n=0}^{N*} \frac{n}{T} \, Pr\{\text{catch of n schools in T}\}. \tag{5.20}$$

This equation can be rewritten as

$$\langle CPUE \rangle = \sum_{n=0}^{N*} \frac{n}{T} \frac{1}{\mathfrak{D}} \binom{n+\nu-1}{n}\left(\frac{\alpha}{\alpha+N(T-nh)}\right)^{\nu}\left(\frac{N(T-nh)}{\alpha+N(T-nh)}\right)^{n}. \tag{5.21}$$

Equation (5.21) involves 4 parameters: T, $\nu$, $\alpha$ and h.  A more complete discussion of (5.21) is found in (26).  Here, one particular case will be studied.  To do this, introduce the following scaled variables

$$\left.\begin{array}{l} M = T/h \\ \hat{\alpha} = \alpha/h = \hat{\alpha}M/T \\ \lambda = N/M \\ \beta = \hat{\alpha}/M = \alpha/T \end{array}\right\} \tag{5.22}$$

The key variables here are M - the maximum possible catch and $\lambda$ - the ratio of abundance to possible catch.  In terms of these scaled parameters, the average CPUE is

$$\langle CPUE \rangle = \frac{\sum\limits_{n=0}^{N*} \frac{n}{T} \binom{n+\nu-1}{n}\left(\frac{\beta}{\beta+\lambda(M-n)}\right)^{\nu}\left(\frac{\lambda(M-n)}{\beta+\lambda(M-n)}\right)^{n}}{\sum\limits_{n=0}^{N*} \binom{k+\nu-1}{k}\left(\frac{\beta}{\beta+\lambda(M-k)}\right)^{\nu}\left(\frac{\lambda(M-k)}{\beta+\lambda(M-k)}\right)^{k}}. \tag{5.23}$$

As before, there are three limits of interest here.  These are: $\lambda \to \infty$, $\lambda \to 0$, and $\lambda \sim 1$.  If $\lambda \to \infty$, which corresponds to $N \to \infty$, it is easily seen that

$$\langle CPUE \rangle \to M/T. \tag{5.24}$$

Thus, for very large abundance one spends all the time fishing and catches, with probability approaching 1, M schools. If $\lambda \to 0$ one finds

$$\langle CPUE \rangle \sim \frac{\nu}{\beta T} \lambda(M-1) \qquad (5.25)$$

so that average CPUE is proportional to population abundance. The third case, $\lambda \sim O(1)$ must be studied numerically. For computations the choice $\beta = \nu = 1$ was made. The choice $\beta = 1$ is consistent with a Poisson (no depletion) model -- i.e. relatively short times. The choice $\nu = 1$ corresponds to considerable uncertainty (a CV of q of 100%). Figure 5.2 shows plots of $\log\{T\langle CPUE\rangle\}$ versus $\log\{\lambda\}$ for two values of M. The slopes are about 0.4, showing that for a range of $\lambda$, $CPUE \sim AN^{.4}$. Extensions and further discussion of this particular model are found in (26).

Both of these models show that when one includes handling and uncertainty in catchability -- both real facets of a fishery -- then reasonable assumptions may lead to $CPUE = AN^P$ with p considerably less than one.


## 6.  Trap Spacing in Agricultural Pest Control

One method for the detection of an invasive, exotic pest -- such as the fruit fly in California -- is by means of trapping (see (33) for a discussion of some problems associated with trapping and detection of fruit flies). The following problem was posed by Dr. D. Chambers, USDA, Gainesville, Florida, regarding the detection of an invasive pest.

Imagine a set of C cities that are potential hosts for such a fruit fly invasion. The cities are separated in space by considerable distances and the region between cities does not support the pest. (This situation applies, for example, to the central valley of California with the cities being Sacramento, Stockton, Modesto, Fresno and Bakersfield). At some point a pest population may be introduced into one of these cities, but one does not know which city nor when the infestation will occur. The detection of the pests, during a time period of length T, occurs by means of traps. There are a total of N traps available. This problem involves a large number of difficult modeling issues. The one of interest for this paper is the following. Consider two possible trapping strategies:

$S_1$:  Leave N/C traps in each city for
the entire period of length T.

$S_2$:  Leave all N traps in one city for
a period of length T/C. Then move
them to another city for T/C, etc.

In both cases, the traps are spread in a regularly spaced grid in the city. The question of interest is this one: Ignoring costs of moving traps, given that a pest is present, which strategy leads to a higher probability of detecting the pest?

For a pest such as the fruit fly, one can assume that the motion of the pest is a simple random walk until the fly enters the region where the trap attractant affects it. Once inside this "trap radius" the fly is immediately drawn to the trap. Imagine traps regularly spaced at a distance $R_s$ from each other and that at time $t$ in the trapping period the radius of the trap is $r(t)$ (this radius might depend upon time, for example, because of ageing or dispersal of the attractant). Imagine a fly starting at the origin and executing a random walk. One needs to calculate the probability of detection by time $T$. The two strategies $S_1$ and $S_2$ now correspond to a trap spacing of $R_s$ and detection time $T$ or spacing of $R_s/C$ and detection time $T/C$ respectively. Let $Z$ denote all of the traps.

Let $f(x,t,Z)$ denote, as in Section (2.6), the joint probability density for the position of the fly at time $t$ and non-trapping. For purposes of simplicity, it is assumed that the fly moves in two dimensions. Then $f(x,t,Z)$ satisfies

$$\frac{\partial f}{\partial t} = \frac{s}{2}\left[\frac{\partial^2 f}{\partial x_1^2} + \frac{\partial^2 f}{\partial x_2^2}\right] - \psi(x,t,Z)f. \tag{6.1}$$

The detection function $\psi(x,t,Z)$ is given as follows. Let $z_i$ denote the center of the $i^{th}$ trap and let

$$H(x,t,z_i) = \begin{cases} 1 & \text{if } \|x-z_i\| < r(t) \\ 0 & \text{otherwise} \end{cases} \tag{6.2}$$

Then assume that

$$\psi(x,t,Z) = \Psi \sum_i H(x,t,z_i) \tag{6.3}$$

where $\Psi$ is a constant. As $\Psi$ increases, the trap becomes more and more effective.

The probability of detection by some time $s$, $P_D(s)$, is then given by

$$P_D(s) = 1 - \iint f(x,s,Z)dx_1 dx_2. \tag{6.4}$$

In order to use (6.1) and (6.4), initial and boundary conditions must be given. For the initial data, take

$$f(x,0,Z) = \frac{1}{2\pi\sigma^2} \exp\left\{-\frac{(x_1^2+x_2^2)}{2\sigma^2}\right\} \tag{6.5}$$

where $\sigma \ll 1$. As $\sigma \to 0$, (6.5) becomes a delta function centered at the origin. For boundary conditions, simply take a decay condition as $x_1$, $x_2 \to \infty$. The integral in (6.4) is simplified somewhat by the following observation. In the absence of search, i.e. with $\psi \equiv 0$, the solution of (6.1) is easily seen to be the Gaussian density

$$\rho(x,y,t) = \frac{1}{2\pi(\sigma^2 + \epsilon t)} \exp\left\{-\frac{(x_1^2 + x_2^2)}{2(\sigma^2 + \epsilon t)}\right\}. \tag{6.6}$$

The level curves of $\rho(x,y,t)$ are circles and

$$\iint\limits_{r < R} \rho(x,y,t)dx\,dy = \int\limits_{r < R} \int_0^{2\pi} \frac{1}{2\pi(\sigma^2 + \epsilon t)} \exp\left\{-\frac{r^2}{2(\sigma^2 + \epsilon t)}\right\} d\theta\,rdr$$

$$= 1 - \exp\left\{-\frac{R^2}{2(\sigma^2 + \epsilon t)}\right\}. \tag{6.7}$$

Thus, a circle that contains the wandering fly with probability $\alpha$ has radius

$$R_\alpha = \sqrt{2(\sigma^2 + \epsilon t)\,|\log(1-\alpha)|}\,. \tag{6.8}$$

By choosing $\alpha$ sufficiently close to 1, the region over which (6.4) needs to be computed can be easily specified.

The solution in (6.1) is obtained by the application of the "ray method" of J. B. Keller (34) to diffusion equations. The application of this method to diffusion equations in general is described in (35,36) and to search equations in particular is described in (21). The method is an asymptotic one, based on the assumption that $\epsilon$ is small. For the problem of fruit flies, the following parameters are appropriate:

| | |
|---|---|
| Search time | 300 - 2600 hours |
| $\sigma$ | $2.5 \times 10^{-5}$ to $2.5 \times 10^{-4}$ mi |
| $\epsilon$ | $1.6 \times 10^{-5}$ to $1.6 \times 10^{-3}$ mi$^2$/hr |
| Trap radius | 100 - 200 feet |
| Trap spacing | .2 - 1.0 mi |

Thus, if one uses a scaling of trap spacing for length and days for time, $\epsilon$ is indeed a small parameter. The asymptotic solution of (6.1) is given by (21)

$$f(x,t,Z) = \frac{1}{2\pi(\epsilon t + \sigma^2)} \exp\left\{-\frac{(x_1^2 + x_2^2)}{2(\epsilon t + \sigma^2)}\right\} \exp\left\{-\int_0^t \psi(x^R(s),s,Z)ds\right\}. \tag{6.9}$$

In (6.9), $x^R(s)$ is the "ray" from the initial density to the point $x$ that reaches $x$ at time $t$. (This undoubtedly sounds somewhat arcane; more complete descriptions are found in (21), (34), (35), (36)). The ray is given by

$$x_1^R(s) = \frac{x_1}{(1+t/\sigma^2)}[1 + s/\sigma^2] \tag{6.10}$$

where $\tilde{\sigma} = \sigma/\sqrt{\epsilon}$ .

It is now a simple matter to numerically compute the integral of $f(x,t,Z)$ over the region of interest. For example, assume that the following parameters are chosen: $\epsilon = 1.6 \times 10^{-5}$ mi$^2$/hr, $\sigma = 2.5 \times 10^{-4}$ mi, trap radius = 100 ft. Then one wants to compare strategies $S_1$ and $S_2$. In particular now assume that

$S_1$: Trap spacing of 1 mile with a
search time of 2520 hrs (15 weeks)

$S_2$: Trap spacing of .2 mile with a
search time of 504 hrs (3 weeks).

Strategy $S_1$ gives a probability of detection of 0.1 and strategy $S_2$ gives a probability of detection of about 0.6. These are not the entire story, however, because one needs to consider the size of the infestation, time at which it was introduced, and other operational factors. The entire problem is too complicated to be considered here, the interested reader is referred to (27).

An interesting heuristic which may help in understanding the results is the following one. Observe from (6.8) that a circle containing the wandering fly with a given level of confidence close to 1 has radius

$$R(t) = k \sqrt{\sigma^2 + \epsilon t} \qquad (6.11)$$

where $k$ is of the order of $5 - 6$ for the confidence level to be of the order of .9999999. Suppose that the trap radius is $r$. Then if one views the problem as placing $n$ discs of area $\pi r^2$ in a disc of size $\pi R(t/n)^2$, the coverage factor $C$ is

$$C = \frac{n\pi r^2}{\pi k^2 (\sigma^2 + \epsilon t/n)} \qquad (6.12)$$

$$= \frac{n^2 r^2}{k^2 (\epsilon t + n\sigma^2)} . \qquad (6.13)$$

Equation (6.13) shows that the coverage factor grows as $n$, unless $\sigma^2 \equiv 0$, in which case it grows as $n^2$. This shows the advantage of the shorter, but more intense trapping period.

## Acknowledgements

This work was partially supported by NSF Grant MCS-81-21659 and by the Agricultural Experiment Station, University of California, Davis.

## References

1. Haywood, K. H. and K. B. Haley (1975), The catching of fish, in Operational Research 75, K. B. Haley (ed.), North Holland, Amsterdam.

2. Mangel, M. (1982), Search effort and catch rates in fisheries, Eur. J. Operational Research, 11:361-366.

3. Mangel, M. and C. W. Clark (1983), Uncertainty, search, and information in fisheries, J. du. Conseil., 41:93-103.

4. Krebs, J. R., A. Kacelnik, and P. Taylor (1978), Test of optimal sampling by foraging great tits, Nature 275:27-31.

5. Clark, C. W. and M. Mangel (1984), Foraging and flocking strategies: information in an uncertain environment, Amer. Natur. 123:626-647.

6. Stefanou, S. (1983), The Optimal Allocation of Scouting Effort and Timing of Pesticide Application, Ph.D. Dissertation, Department of Agricultural Economics, University of California, Davis.

7. Stefanou, S., Mangel, M. and J. E. Wilen (1985), Pest scouting, value of information, and optimal spraying decisions, J. Ag. Econ., in press.

8. Plant, R. E. and T. Wilson (1985), A Bayesian method for sequential sampling and forecasting in agricultural pest management, Biometrics, to appear.

9. Koopman, B. O. (1980), Search and Screening, Pergamon Press, N.Y.

10. Neyman, J. (1949), On the problem of estimating the number of schools of fish, Calif. Univ. Pub. Stat. 1:21-36.

11. Feller, W. (1968), An Introduction to Probability Theory and Its Applications, Vol. I, Wiley, New York.

12. Karlin, S. and H. M. Taylor (1981), A Second Course in Stochastic Processes, Academic Press, New York.

13. Feller, W. (1971), An Introduction to Probability Theory and Its Applications, Vol. II, Wiley, N.Y.

14. Abramowitz, M. and I. Stegun (1964), Handbook of Mathematical Functions, Nat. Bur. of Standards, Washington, D.C.

15. Mangel, M. and J. H. Beder (1985), Search and Stock Depletion: Theory and Applications, Can. J. Fish Aq. Sci., 42:150-163.

16. Reed, W. (1984), A method of analyzing catch-effort data which allows for randomness in the catching process, preprint, Department of Mathematics, University of Victoria, British Columbia, Canada.

17. Koopman, B. O. (1980), Search and Screening, Pergamon Press, N.Y.

18. Stone, L. D. (1975), Theory of Optimal Search, Academic Press, N.Y.

19. Washburn, A. R. (1981), Search and Detection, Mil. Applic. Section, ORSA, Arlington, Va.

20. Riesz, F. and B. Sz-Nagy (1978), Functional Analysis, F. Ungar, New York.

21. Mangel, M. (1985), Search Theory, Springer Lecture Notes in Control, to appear.

22. Mangel, M. (1981), Search for a randomly moving object, SIAM J. Appl. Math. 40:327-338.

23. Hellman, O. (1970), On the effect of search upon the probability distribution of a target whose motion is a diffusion process, Ann. Math. Stat. 41:1717-1724.

24. Schuss, Z. (1980), Theory and Applications of Stochastic Differential Equations, Wiley, N.Y.

25. U.S. Coast Guard (1973), National Search and Rescue Manual, CG 308, Superintendent of Documents, Washington, D.C.

26. Mangel, M. (1985), On the relationship between catch per unit effort and stock abundance, Ecol. Modelling, to appear.

27. Mangel, M. (1985), On the determinants of trap spacing, to appear in Proc. NATO ARW, "Pest Control: Operations and Systems Analysis in Fruit Fly Management", M. Mangel (ed.).

28. Murphy, G. I. (1977), Clupeoids, Chapter 12 in Fish Population Dynamics, J. Gulland (ed.), Wiley, New York.

29. Ulltang, O. (1980), Factors affecting the reaction of pelagic fish stocks to exploitation and requiring a new approach to assessment and management, Rapp P -V. Reun. Cons. Int. Explor. Mer. 177:489-504.

30. Paloheimo, F. E. and L. M. Dickie (1964), Abundance and fishing success, Rapp et. P-V. Cons. Reun. int. Explor. Mer. 155:152-163.

31. Beddington, J. R. (1979), On some problems of estimating population abundance from catch data, Rept. Int. Whale Comm. 29:149-154.

32. Cooke, J. (1985), On the relationship between catch per unit effort and whale abundance, Rep. Int. Whal. Comm., to appear.

33. Mangel, M., R. E. Plant, and J. R. Carey (1984), Rapid delimiting of pest infestations: A case study of the Mediterranean fruit fly, J. Appl. Ecol. 21:563-579.

34. Keller, J. B. (1978), Rays, waves, and asymptotics, Bull. Amer. Math. Soc. 84:727-750.

35. Cohen, J. and R. M. Lewis (1967), A ray method for the asymptotic solution of the diffusion equation, J. Inst. Math. Applic., 3:266-290.

36. Ludwig, D. (1975), Persistence of dynamical systems under random perturbations, SIAM Review 17:605-640.

Postscript: I recently discovered that F. Anscombe scooped me on section 2.5 (transformations to normality) in 1948! (Biometrika 35:242-245).

# Biomathematics

Managing Editor: S.A.Levin

Editorial Board: M.Arbib, H.J.Bremermann, J.Cowan, W.M.Hirsch, S.Karlin, J.Keller, K.Krickeberg, R.C.Lewontin, R.M.May, J.D.Murray, A.Perelson, L.A.Segel

Volume 14
## C.J.Mode
## Stochastic Processes in Demography and Their Computer Implementation

1985. 49 figures, 80 tables. XVII, 389 pages. ISBN 3-540-13622-3

**Contents:** Fecundability. – Human Survivorship. – Theories of Competing Risks and Multiple Decrement Life Tables. – Models of Maternity Histories and Age-Specific Birth Rates. – A Computer Software Design Implementing Models of Maternity Histories. – Age-Dependent Models of Maternity Histories Based on Data Analyses. – Population Projection Methodology Based on Stochastic Population Processes. – Author Index. – Subject Index.

Volume 13
## J.Impagliazzo
## Deterministic Aspects of Mathematical Demography

**An Investigation of the Stable Theory of Population including an Analysis of the Population Statistics of Denmark**

1985. 52 figures. XI, 186 pages. ISBN 3-540-13616-9

**Contents:** The Development of Mathematical Demography. – An Overview of the Stable Theory of Population. – The Discrete Time Recurrence Model. – The Continuous Time Model. – The Discrete Time Matrix Model. – Comparative Aspects of Stable Population Models. – Extensions of Stable Population Theory. – The Kingdom of Denmark – A Demographic Example. – Appendix. – References. – Subject Index.

Volume 12
## R.Gittins
## Canonical Analysis

**A Review with Applications in Ecology**

1985. 16 figures. XVI, 351 pages. ISBN 3-540-13617-7

**Contents:** Introduction. – **Theory**: Canonical correlations and canonical variates. Extensions and generalizations. Canonical variate analysis. Dual scaling. – **Applications**: General introduction. Experiment 1: an investigation of spatial variation. Experiment 2: soil-species relationships in a limestone grassland community. Soil-vegetation relationships in a lowland tropical rain forest. Dynamic status of a lowland tropical rain forest. The structure of grassland vegetation in Anglesey, North Wales. The nitrogen nutrition of eight grass species. Herbivore-environment relationships in the Rwenzori National Park, Uganda. – **Appraisal and Prospect**: Applications: assessment and conclusions. Research issues and future developments. – **Appendices**: Multivariate regression. Data sets used in worked applications. Species composition of a limestone grassland community. – References. – Species' index. – Author index. – Subject index.

Springer-Verlag
Berlin
Heidelberg
New York
Tokyo

# Journal of

# Mathematical Biology

ISSN 0303-6812                                    Title No. 285

**Editorial Board:**

**H. T. Banks,** Providence, RI; **J. D. Cowan,** Chicago, IL;
**J. Gani,** Lexington, KY; **K. P. Hadeler** (Managing Editor),
Tübingen; **F. C. Hoppensteadt,** Salt Lake City, UT;
S. A. Levin (Managing Editor), Ithaca, NY; **D. Ludwig,**
Vancouver; **L. J. D. Murray,** Oxford, **L. T. Nagylaki,**
Chicago, IL; **L. A. Segel,** Rehovot
in cooperation with a distinguished advisory board.

For mathematicians and biologists working in a wide spectrum
of fields, the **Journal of Mathematical Biology** publishes:

- papers in which mathematics in used to better understand
  biological phenomena
- mathematical papers inspired by biological research and
- papers which yield new experimental data bearing on mathematical models.

Contributions also discuss related areas of medicine, chemistry,
and physics.

*Articles from a recent issue:*

**E. Doedel:** The computer-aided bifurcation analysis of
predator-prey models
**S. Karlin, S. Lessard:** On the optimal sex-ratio: A stability
analysis based on a characterization for one-locus multiallele
viability models
**J. M. Mahaffy, C. V. Pao:** Models of genetic control by repression
with time delays and spatial effects
**P. Creegan, R. Lui:** Some remarks about the wave speed and
traveling wave solutiions of a nonlinear integral operator
**H. Aargaard-Hansen, G. F. Yeo:** A stochastic discrete generation
birth, continuous death population growth model and its
approximate solution
**F. M. Hoppe:** Pólya-like urns and the Ewens' sampling formula
**M. Weiss:** A note on the rôle of generalized inverse Gaussian
distributions of circulatory transit times in pharmacokinetics
**R. Dal Passo, P. de Mottoni:** Aggregative effects for a reaction-
advection equation.

Subscription information and sample copy upon request

Springer-Verlag
Berlin
Heidelberg
New York
Tokyo